National Symposium on Family Issues

Volume 12

Each year, the Population Research Institute and Social Science Research Institute at Penn State hold a two-day symposium that focuses on a key problem of relevance to family studies. The symposium, known as the *National Symposium on Family Issues*, brings together 200 or more scholars, practitioners, and policy experts to:

- Promote interdisciplinary dialogue to stimulate research on family issues.
- Advance scholarly excellence by inviting leaders in the field to present their work.
- Identify important issues that do not receive sufficient attention from researchers who study families.

The symposium organizers – Professors Susan McHale, Valarie King, and Jennifer E. Glick – work to connect family scholars from diverse fields, such as demography, sociology, human development, psychology, education, economics, anthropology, law, and history. The *National Symposium on Family Issues* is a landmark event in ongoing efforts to understand more thoroughly the challenges facing contemporary families.

At each annual symposium, nine or more leading scholars are convened to present and critique research on a focal topic and discuss implications for effective programs and policies for families. Books based on each symposium provide additional background and detail about current research and applications for evidence based policy and programs, allowing the research to reach a wider audience and impact the national conversation. Past volumes have received favorable reviews and are used as reference works by researchers, professors, students as well as clinicians and other professionals.

Selena E. Ortiz • Susan M. McHale
Valarie King • Jennifer E. Glick
Editors

Environmental Impacts on Families

Change, Challenge, and Adaptation

Springer

Editors
Selena E. Ortiz
Department of Health Policy and Administration
Penn State
University Park, PA, USA

Susan M. McHale
Department of Human Development and Family Studies
Penn State
University Park, PA, USA

Valarie King
Department of Sociology
Penn State
University Park, PA, USA

Jennifer E. Glick
Department of Sociology
Penn State
University Park, PA, USA

ISSN 2192-9157 ISSN 2192-9165 (electronic)
National Symposium on Family Issues
ISBN 978-3-031-22648-9 ISBN 978-3-031-22649-6 (eBook)
https://doi.org/10.1007/978-3-031-22649-6

This Springer imprint is published by the registered company Springer Nature Switzerland AG
The registered company address is: Gewerbestrasse 11, 6330 Cham, Switzerland

Acknowledgments

The editors gratefully acknowledge essential core support provided by the Eunice Kennedy Shriver National Institute of Child Health and Human Development (NICHD; R13-HD048150), as well as ongoing advice from Regina Bures at NICHD. We also thank our Penn State sponsors, the Social Science Research Institute (SSRI) and Population Research Institute (PRI). Our sponsors' support enabled us to attract outstanding scholars from a range of backgrounds and disciplines from the community of researchers, practitioners, and policy makers on whom the quality of the series depends.

The efforts of many went into planning the 2021 symposium and producing this volume. We thank our Internal Advisory Board, Professors Sarah Damaske, Greg Fosco, Alyssa Gamaldo, Yolanda Jackson, Nancy Luke, Mary Shenk, Doug Teti, Ashton Verdery, and Dawn Witherspoon for their continued help in developing the foci of our annual symposium. Professors Alexis Santos, Heather Randell, and Selena Ortiz served as moderators of the three sessions of the 2021 symposium. We also thank Kristie Auman-Bauer in SSRI for her help in publicizing the symposium. The virtual family symposium, necessitated by the COVID pandemic, was skillfully orchestrated by Mark Hixon and Russell Houtz from the SSRI/PRI IT Core. Finally, the symposium and book would not have been possible without Carolyn Scott's organizational skills, commitment, and attention to the many details that go into developing an engaging conference and producing a scholarly volume.

Selena E. Ortiz
Susan M. McHale
Valarie King
Jennifer E. Glick

About the Book

Families are embedded in larger contexts that have implications for family relationships and well-being. A large body of research by family scholars has documented multiple dimensions of the broader social environment – including cultural, economic, and political contexts – in family functioning. Family scholars have paid less attention, however, to how dimensions of the physical environment may impact families, including factors that impede health and healthy behaviors, as well as factors that may protect families and promote their resilience in the face of change and challenge. Toward promoting equity, significant efforts must be directed at recognizing and addressing the physical conditions – and the structural and systemic factors that allow these physical conditions to persist – that undermine the health and well-being of children and families. To that end, this volume highlights discussion and emerging evidence that reinforces the call for significant and sustainable investments that improve the physical environment in ways that support families.

Toward stimulating novel interdisciplinary and translational research on families, the 2021 National Symposium on Family Issues showcased the work of teams of scholars focusing on the role of the physical environment in family relationships, behaviors, and well-being. The symposium focused on three critical dimensions: disasters, climate change, and the built environment.

The chapters in the first section address the social, demographic, and health impacts of disasters such as hurricanes and tsunamis, on communities, families, and children. Rich data sources and mixed-method techniques are utilized to produce in-depth appreciations of the multiple and complex ways in which disasters impact families, both in the immediate aftermath and over the long term. The results from these analyses underscore the importance of understanding the sustained impacts of disasters on the health and well-being of families, as well as emphasize the role of individual and community resilience in long-term recovery from disasters. The authors also suggest ways in which evidence-based prevention and treatment interventions and government programs and policies can assist families and family members to recover after a disaster. However, in light of the critical role of family in post-disaster recovery, the authors acknowledge the dearth of family-focused programming and recommend providing substantial support for family-oriented intervention research and development.

The second section focuses on climate change and the intersection of environmental conditions, socioeconomic disparities, and community resilience that has shaped new realities and constraints for families. The chapters in this section advance understandings of how environmental change impacts maternal and child health, family functioning, and adaptation across diverse communities around the world. In their research, authors consider the numerous ways in which climate change has altered family life and opportunities, such as when and why families pursue migration, as well as issues of environmental justice. Recognizing the range of interconnected factors – demographic, social, economic, political, and environmental – that independently and collectively function to increase families' susceptibility to climate change-induced conditions, the authors argue that novel approaches to examining the processes that link these conditions to health and demographic outcomes are essential. Such innovative approaches require the development and application of new theory, data, and analytic tools, as well as the integration of fields of scholarship to maximize insights on how climate change impacts families.

The third section considers the ways in which dimensions of the built environment – from the proximal environment of homes to neighborhood and larger community environments – have effects on the health and well-being of children, adolescents, and their families. The chapters in this section examine the ways in which features of home environments, as well as extensions of home environments (e.g., afterschool settings, nearby play centers, etc.), *get under the skin* to affect family processes and youth physical and mental health; how food access and opportunities for physical activity in communities serve to promote youth and family health; and how neighborhood development can reduce burdens experienced by family caregivers who depend on and must navigate environmental characteristics. The authors consider the efficacy of adopting community-driven and systems thinking approaches to create and maintain opportunities for physical activity in low-resourced communities and to improve the urban conditions that families must negotiate to provide care for their loved ones. Future investigations focusing on how the built environment impacts children of color, as well as immigrant, refugee, unhoused, and otherwise displaced children, are also recommended.

The research presented in this volume, as discussed in its final chapter, incorporates diverse theoretical frameworks, engages interdisciplinary perspectives, and applies robust methodologies to investigate how the physical environment – disasters, climate change, and the built environment – can exact severe harms upon family functioning, health, and well-being. The 2021 National Symposium on Family Issues generated vital exchange on the significance of designing, implementing, and evaluating efforts that can protect against these harms. Taken together, the research described in this volume offers opportunities to reflect upon the breadth of the constraints and challenges families encounter in their physical environments – a first step toward conducting novel research to serve as the foundation for developing and implementing policies and programs that build resiliency among families and, ultimately, achieve health equities.

Selena E. Ortiz
Susan M. McHale
Valarie King
Jennifer E. Glick

Contents

About the Editors

Selena E. Ortiz, Ph.D., is Associate Professor of Health Policy and Administration at Penn State.

Susan M. McHale, Ph.D., is Distinguished Emeritus Professor of Human Development and Family Studies and Emeritus Professor of Demography at Penn State.

Valarie King, Ph.D., is Professor of Sociology, Demography, and Human Development and an Associate in the Population Research Institute at Penn State.

Jennifer E. Glick, Ph.D., is Arnold S. and Betty G. Hoffman Professor in Sociology at Penn State.

Contributors

Mariana C. Arcaya Department of Urban Studies and Planning, Massachusetts Institute of Technology, Cambridge, MA, USA

Andrew Binet Department of Urban Studies and Planning, Massachusetts Institute of Technology, Cambridge, MA, USA

Amanda R. Carrico University of Colorado, Boulder, CO, USA

Audrey Dorélien Hubert H. Humphrey School of Public Affairs, University of Minnesota, Minneapolis, MN, USA

Gary W. Evans College of Human Ecology, Cornell University, Ithaca, NY, USA

Kim T. Ferguson Sarah Lawrence College, Bronxville, NY, USA

Elizabeth Frankenberg University of North Carolina at Chapel Hill, Chapel Hill, NC, USA

Kathryn Grace Department of Geography, Environment, and Society, University of Minnesota, Minneapolis, MN, USA

Nicholas Ingwersen Carolina Population Center, University of North Carolina at Chapel Hill, Chapel Hill, NC, USA

Rene Iwo Department of Sociology, University of North Carolina at Chapel Hill, Chapel Hill, NC, USA

Jorden E. Jackson Department of Sociology, Penn State, University Park, PA, USA

Herman Jenkins Friends of the Rouge, Plymouth, MI, USA

Laurie Lachance School of Public Health, Department of Health Behavior and Health Education, The University of Michigan, Ann Arbor, MI, USA

Devin Malloy McCauley Department of Pediatrics, Stanford University School of Medicine, Stanford, CA, USA

Jenna M. Muller School of Social Work, University of Illinois at Urbana-Champaign, Urbana, IL, USA

Tara Powell School of Social Work, University of Illinois at Urbana-Champaign, Urbana, IL, USA

Martha Quinn School of Public Health, Department of Health Behavior and Health Education, The University of Michigan, Ann Arbor, MI, USA

Samuel Shopinski The National Kidney Foundation of Michigan, Ann Arbor, MI, USA

Cecep Sumantri SurveyMETER, Yogyakarta, Indonesia

Theresa Tejada The National Kidney Foundation of Michigan, Ann Arbor, MI, USA

Duncan Thomas Department of Economics, Duke University, Durham, NC, USA

Greta Wetzel Save the Children, Arlington, VA, USA

Part I
Environmental Disasters

Chapter 1
Impacts of Disaster-Induced Death and Destruction on Health and Mortality Over the Longer Term

Elizabeth Frankenberg, Nicholas Ingwersen, Rene Iwo, Cecep Sumantri, and Duncan Thomas

Climate change is increasing the frequency and intensifying the force of natural disasters at the same time that populations in vulnerable areas are growing in size. Projections that take the combination of these forces into account indicate that relative to their parents and grandparents, today's children and young adults will experience a four- to sevenfold increase in the number of extreme events they live through (Thiery et al., 2021). Understanding the sustained impacts of these events on health and well-being is critically important, but a key constraint is the paucity of high-quality longitudinal data that can advance the science.

In this chapter, we use data from an extremely rich population-representative longitudinal survey, the Study of the Tsunami Aftermath and Recovery (STAR), to explore how both longer-term survival and psychosocial health of individuals who experienced a natural disaster are affected by various types of exposure in the 15 years after the event, in comparison to individuals who were not directly exposed. We study the 2004 Indian Ocean earthquake and tsunami. The disaster, which killed

---

E. Frankenberg (✉) N. Ingwersen
Carolina Population Center, University of North Carolina at Chapel Hill, Chapel Hill, NC, USA
e-mail: e.frankenberg@unc.edu; ningwersen@unc.edu

R. Iwo
Department of Sociology, University of North Carolina at Chapel Hill, Chapel Hill, NC, USA
e-mail: rene.iwo@unc.edu

C. Sumantri
SurveyMETER, Yogyakarta, Indonesia
e-mail: sumantri@surveymeter.org

D. Thomas
Department of Economics, Duke University, Durham, NC, USA
e-mail: dthomas@econ.duke.edu

S. E. Ortiz et al. (eds.), *Environmental Impacts on Families*, National Symposium on Family Issues 12, https://doi.org/10.1007/978-3-031-22649-6_1

an estimated quarter of a million people, is one of the most devastating natural disasters in recorded history. Nowhere was hit harder than coastal Aceh, Indonesia. The tsunami completely destroyed some communities but left other comparable communities untouched. STAR is uniquely well-suited for this research: the baseline was conducted 10 months before the tsunami as part of Statistics Indonesia's annual household survey program. We have followed survivors for 15 years post-tsunami. The tsunami was completely unanticipated, and the location of the communities that sustained damage is a complex function of the location of the precipitating earthquake and the topography of the sea floor and coastline. Leveraging the natural experiment of the tsunami, we provide evidence that credibly identifies the causal impact of the disaster on mortality and psychosocial health over the longer term.

Context of the Disaster

Early in the morning on Sunday, December 26, 2004, one of the most powerful earthquakes in recorded history occurred some 150 miles from the coast of the island of Sumatra, Indonesia. The earthquake displaced a trillion tons of water, which formed a series of tsunami waves that hit the northern coast of Sumatra about 15 min later and eventually reached across the entire Indian Ocean (Rubin et al., 2017). The tsunami was completely unexpected. Geological records indicate that the last tsunami to hit mainland Sumatra was over 600 years ago (Monecke et al., 2008).

Aceh, the northern most province on the island of Sumatra, was hardest hit. Along 800 km of the coast, communities experienced varying degrees of inundation, resulting in destruction of the built and natural environment and the deaths of more than 170,000 people (Lavigne et al., 2009).

Impacts varied considerably even between areas quite close to one another. The water's height and inland reach were a function of slope, water depth, and coastal topography (Ramakrishnan et al., 2005). Along parts of the west coast of Aceh, trees up to 13 m tall lost their bark (Borrero, 2005). At the beachfront in Banda Aceh, the province's capital and largest city, the water was as deep as 9 m, though rarely exceeded the height of a two-story building (Borrero, 2005). Low-lying communities within a few kilometers of the coast were largely destroyed, and many of their residents perished. River basins allowed the waves to move inland as much as 9 km in some areas, whereas in other locations they encroached only 3–4 km (Kohl et al., 2005; Umitsu et al., 2007). Areas sheltered by altitude, distance from the coast, or other topographical features sustained damage to structures and deposition of sediment and debris, but larger proportions of the population survived. For some communities the tsunami had few if any direct effects, although the earthquake was felt throughout Aceh and damaged property and infrastructure in some areas that the water never reached. The tsunami affected the transportation network along the coast, and some communities were cut off from the main roads connecting major population centers. In some cases, residents of communities that were not directly impacted by the tsunami saw increased demand for their goods and services, particularly food and housing.

Data

The STAR baseline consists of respondents who participated in a large, population-representative socioeconomic survey (SUSENAS) conducted by Statistics Indonesia in February/March 2004, 10 months before the tsunami. SUSENAS is representative at the *kabupaten* (regency) level. We worked with Statistics Indonesia to select all 11 districts in the province of Aceh that had coastlines which were potentially vulnerable to inundation by a tsunami. Within each selected district, we included all SUSENAS enumeration areas, regardless of distance from the coast. All members of all households enumerated in these districts in the 2004 SUSENAS form the STAR baseline study population.

SUSENAS, a long-standing government survey that is well-known in Indonesia, achieves participation rates that exceed 97%. The survey, which most closely parallels a combination of the Consumer Expenditure Survey and Current Population Survey in the United States, collects information on demographic and socioeconomic characteristics of household members from a key household member. The first STAR follow-up survey took place between May 2005 and July 2006. Four annual follow-ups were conducted thereafter, with additional follow-ups roughly 10 and 15 years after the tsunami.

We triangulated across multiple sources of information to establish survival status for 99% of the baseline (pre-tsunami) respondents. Information came from interviews with household and family members (whose reports we consider most reliable), community leaders, and neighbors. Information from the latter two sources was critical for households in which no members could be located. In each follow-up, every household member was interviewed. Parents or caregivers provided information about children aged 11 years or younger; proxy respondents provided information for adults unable to answer for themselves. The first two follow-up surveys collected detailed information on experiences at the time of the tsunami from each respondent. All surveys included questions on physical health, psychosocial well-being, and behavioral responses to the event, including displacement and migration, as well as information about individual and household demographics and socioeconomic status.

Measuring Exposure to the Tsunami

We investigate links between multiple indicators of exposure to the tsunami and two post-tsunami measures of health—mortality for those who survived the tsunami and post-traumatic stress reactivity (PTSR). Frankenberg et al. (2011) describe mortality at the time of the tsunami, and Frankenberg et al. (2008) describe PTSR in the year after the tsunami. See, also, Ho et al. (2017) and Frankenberg et al. (2020) for studies of mortality. This research builds on those studies. Our data on mortality are derived from our household rosters which we update at each wave to track survival status and movement across locations (and of household members across households).

Following the literature, we summarize the impact of the stressors using an index of the incidence and severity of symptoms of PTSR, based on seven items from the PTSD Checklist Civilian Version (Weathers et al., 1993). For example, respondents were asked whether they have had "repeated, disturbing memories, thoughts, dreams or relived experiences of the tsunami" and "felt very upset when something reminded you of the tsunami." If the respondent did experience the feelings, they were asked, for the period when the feelings were most severe, whether they felt them rarely (coded as 1), sometimes (coded as 2), or often (coded as 3). Respondents who did not experience the feelings are coded as 0. Summing the responses to each of the seven items creates a 21-point scale. Elevated PTSR, a score above 11, is represented by a 1; a 0 represents respondents who do not have elevated PTSR.

These questions were included in the first post-tsunami survey conducted 5 to 16 months after the tsunami, except for a small fraction of respondents (less than 3%) whose first post-tsunami interview took place during the second follow-up at 18 to 24 months after the tsunami.

Exposure to the tsunami was measured at both the community and individual level. At the community level, two classes of measures of exposure are operationalized. Our first measure is an indicator of exposure based on the geographic location of the community where each respondent resided at the time of the tsunami. This measure, in recognition that characteristics of the tsunami wave and coastline topography were key determinants of death and destruction at the time of the tsunami, combines information on that community's elevation above sea level, proximity to the coastline, and tsunami wave height at the closest coastal point to the community. In the analyses this indicator allows us to distinguish respondents who were living in communities that were directly affected by the 2004 tsunami ("tsunami-affected") from respondents who were living in communities at similar risk of exposure to a tsunami but were not directly affected by the 2004 tsunami ("other"). Our second measure of exposure at the community level is the percentage of baseline respondents in the community who died in the tsunami. It was designed to reflect intensity of exposure and varies from no deaths to a staggering three-quarters of the community residents perishing in the tsunami.

Turning to individual-level measures, we asked each surviving respondent about their own experiences of the tsunami. The first set of individual-specific measures reflects experiences that may generate a sense of helplessness or horror (which have been linked with symptoms of post-traumatic stress; Dalgleish, 1999). Specifically, we asked whether the respondent was caught up in the water, was injured at the time of the tsunami, or watched friends or family struggle or disappear in the waves. Any affirmative answer is classified as direct exposure. We also constructed exposure measures that capture loss of family: whether the tsunami killed an individual's spouse, or whether it killed an individual's parent, sibling, or child (regardless of whether the family member was co-residential). These individual-specific indicators of exposure complement the community-level measure of damage, providing more fine-grained indicators of tsunami-related stresses experienced by the respondent. As with PTSR, the individual-specific questions were asked at the first post-tsunami follow-up except for those respondents whose first post-tsunami individual interview took place during the second follow-up.

Finally, to address post-tsunami displacement during the 24 months after the disaster, we developed an indicator identifying respondents who lived in temporary housing during this period: a tent, camp, or barracks. This measure draws on data collected in the first three annual post-tsunami follow-ups.

Our community-level and individual-level direct exposure measures are plausibly exogenous because they depend primarily on characteristics that were outside the control of the respondents at the time of the tsunami. While the exact locations of tsunami impact are reasonably treated as random, residential location is a choice, and it is possible that those who were living in areas that were inundated are different from those who were living elsewhere. To address this concern, we examined the effects of individual-level exposures, drawing contrasts between individuals who, at the time of the tsunami, were living in the same community so that the estimates are not contaminated by differences across communities in vulnerability, socioeconomic status, and the availability of resources.

We focus on mortality and psychosocial health in the 15 years after the tsunami among 5927 individuals from 334 baseline communities who were age 35 and older at the time of the tsunami and who survived to the first post-tsunami interview. Face-to-face interviews were completed with 97.8% of this group in the 15 years after the disaster (0.5% refused, 1.7% were not found). This re-interview rate is unprecedented for a large-scale population-representative follow-up 15 years after baseline and stands out given the extent of displacement and the complexity of conducting fieldwork in the aftermath of the tsunami. It reflects the combination of well-designed and extensively tested tracking protocols, high-quality fieldwork, and the commitment of respondents, enumerators, and team supervisors to the scientific goals of the project.

Methods

We examine the correlates of mortality and levels of PTSR after the tsunami by estimating a sequence of models that allow us to consider different measures of exposure and take into account unobserved factors specific to the community by drawing comparisons among survivors who were living in the same community at the time of the tsunami. For the mortality model, we analyze a binary dependent variable, θ_{ic}, which takes the value 1 if the tsunami survivor, i, who was living in community, c, at baseline died during the 15 years after the tsunami and 0 if the individual survived for 15 years:

$$\theta_{ic} = \alpha + \beta T_c + \gamma X_{ic} + \varepsilon_{ic} \tag{1.1}$$

where T_c indicates community-level tsunami exposure, specifically either (1) whether the respondent's pre-tsunami community was affected by the tsunami (our geographically based measure of exposure described above which parallels exposure measures in most other empirical work on this topic) or (2) the percentage of baseline respondents from the respondent's pre-tsunami community who were killed in the tsunami (which we know from updating the baseline data with survival status at the first follow-up). The vector X_{ic} includes individual background characteristics measured at the pre-tsunami baseline: age (in years), education, whether the respondent was married, and household expenditures per capita (a well-established measure of economic resources; Deaton, 1997). We also include a measure of height (measured in the follow-up surveys) as a control for health endowment. Unobserved heterogeneity is captured by ε_{ic}.

The baseline model is extended to examine how individual exposures to the disaster's direct impacts are related to death over the next 15 years:

$$\theta_{ic} = \alpha + \beta T_c + \lambda E_{ic} + \gamma X_{ic} + \varepsilon_{ic} \tag{1.2}$$

where E_{ic} is a vector of measures of exposure based on individual reports of experiences and losses at the time of the tsunami, whether the respondent experienced high levels of post-traumatic stress reactivity, and whether the respondent lived in temporary housing in the 2 years after the tsunami.

To address the possibility that the community-level measures of tsunami exposure reflect factors such as vulnerability, and to highlight the role of individual exposures, the model is extended to draw comparisons in variation in exposure between individuals within the same community:

$$\theta_{ic} = \alpha + \lambda E_{ic} + \gamma X_{ic} + \mu_c + \varepsilon_{ic} \tag{1.3}$$

where μ_c represents enumeration area (EA) indicators that absorb the influence of all community-level variation that does not change over time and affects mortality in a linear and additive way. This includes levels of vulnerability that are shared by community members, the extent of damage in the community because of the earthquake and tsunami, post-tsunami reconstruction, and pre-tsunami levels of infrastructure and economic activity, as well as other unobserved community-level factors that might be correlated with both choice of pre-tsunami location and mortality.

When we shift to post-traumatic stress reactivity, our dependent variable varies from 0 to 21, with higher scores corresponding to a combination of higher incidence and greater intensity of symptoms. For PTSR, we estimate models 2 and 3 to examine how the exposure measures at the time of the tsunami and in the 2 years after the disaster relate to psychosocial health at 5, 10, and 15 years post-disaster.

Results: Surviving the Tsunami

We begin by presenting descriptive statistics for key variables (Table 1.1). Our respondents were living in 334 communities in 2004 before the tsunami, of which 191 were directly affected by the tsunami, as indicated by our geographically based measure of exposure at the community level (Panel A). This dichotomy captures tsunami-related exposures well and separates areas where there was tsunami-related mortality (among household members at baseline, an average of 14.3% were killed) from areas where death due to the tsunami was negligible (less than 1%, on average, were killed).

We use this dichotomous measure to contrast the percentage of baseline respondents who died during the tsunami and the percentage among those who survived the disaster who died in the next 15 years (Fig. 1.1).

The first series of four bars presents the results for the full sample of 6687 individuals in the baseline (to estimate mortality between baseline and the first follow-up) and the 5927 baseline individuals who survived to the first follow-up (to estimate mortality between the first follow-up and the 15-year follow-up). All were 35 years of age or older at the time of the baseline survey in 2004. Not surprisingly, among those from tsunami-affected communities (in red), mortality was markedly higher (17%) than for those from other communities (in blue, 0.6%). However, in the 15 years after the disaster, the direction of this differential reverses: 27% of those from "other" communities are dead by the 2020 survey, whereas only 23% of those from tsunami-affected communities have died.

This reversal is consistent with the idea that the tsunami exerted a force of positive selection—causing the deaths of more frail members of communities in which the waves came ashore and left behind a group of survivors who were, on average, more robust than individuals in the communities where waves did not strike. Evidence for positive mortality selection emerged in the first 5 and 10 years after the tsunami (Ho et al., 2017; Frankenberg et al., 2020).

Age and sex were important determinants of survival during the tsunami (Frankenberg et al., 2011) and therefore may shape survival patterns in its aftermath. Accordingly, we also present mortality patterns for four groups differentiated

Table 1.1 Descriptive statistics

			Communities directly affected by the tsunami	Other communities
A.	Community exposures	Number of communities	191	143
		Percent died at time of tsunami	14.3	0.4
B.	Individual tsunami exposures (%)	Direct experience with waves	20.4	4.3
		Death of spouse	5.5	2.3
		Death of parent, child, and/or sibling	22.1	9.4
		In temporary housing after tsunami	20.8	6.4
		High post-traumatic stress reactivity	21.2	14.1
C.	Baseline characteristics	Male (%)	50.7	48.3
		Age at time of tsunami (mean in years)	49.7	50.9
		Years of education (mean)	7.7	6.7
		Widowed or divorced (%)	15.4	14.1
		Ln of per capita expenditure pre-tsunami	12.8	12.8
	N		3185	2742

Note: Individual exposures and baseline characteristics are computed for tsunami survivors age 35 and older at time of tsunami. All estimates weighted to represent population of survivors at time of tsunami

by sex and by age (respondents aged 35–49 years at the tsunami are distinguished from individuals aged 50 years and older). The evidence for positive selection is replicated for each of these four groups, although it is particularly strong for older males.

In the aggregate, and without controls for other factors, mortality selection appears to be positive. We explore this result in several ways: by examining mortality after the tsunami as a function of the community-level mortality rate during the disaster (to better distinguish the magnitude of the tsunami's impact), by controlling for baseline demographic and socioeconomic characteristics, and by including measures of individual-level exposures. Descriptive statistics for these variables are presented in panels B and C of Table 1.1.

With respect to the individual measures of tsunami exposure, for all but PTSR the percentage exposed is at least twice as high for respondents originally from directly affected communities as for respondents from other communities. With respect to demographic and socioeconomic characteristics at baseline, survivors in the directly affected and other communities are very similar. Individuals from

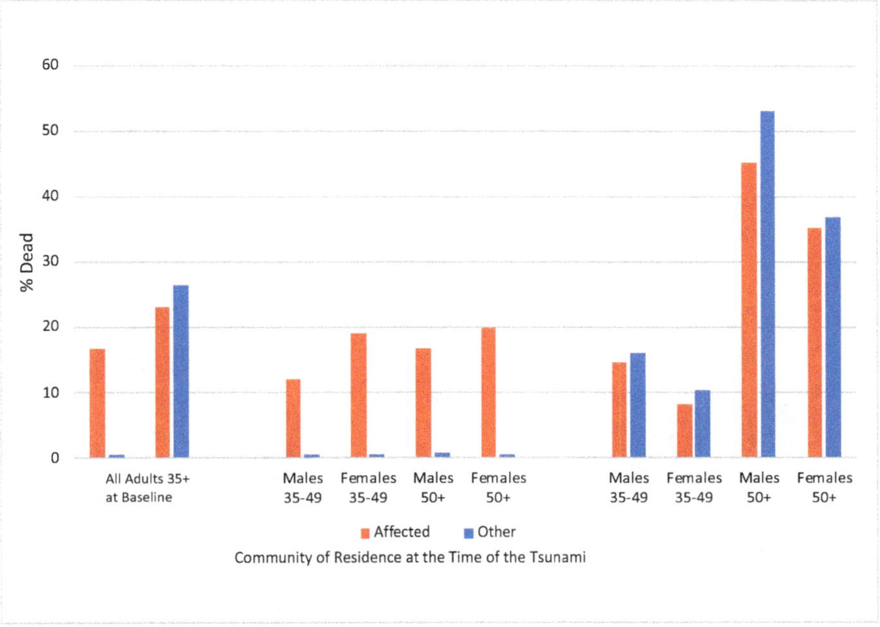

Fig. 1.1 Percent dead at and in the 15 years after the tsunami by community-level tsunami exposure, age, and sex

directly affected communities are a little more likely to be male, are about a year younger, have an additional year of education, and are more likely to be widowed. These differences largely reflect differential tsunami survival. In contrast, there is no difference between the directly affected and other communities in the level of household resources at the time of the tsunami, as indicated by the logarithm of household per capita expenditure measured in the pre-tsunami baseline, which is widely considered to be the best indicator of resource availability in low-income settings and is a good time-varying marker of socioeconomic status (SES).

Mortality in the 15 Years After the Tsunami

In Table 1.2 we present the results from Model 1 (columns 1 and 2), to explore the relationship between community-level measures of exposure and mortality in the 15 years after the tsunami. The dependent variable is multiplied by 100 so that the coefficients are interpreted as changes in percentage points. The coefficients associated with both the dichotomous and the continuous measures of community-level tsunami impact are negative and statistically significant. Both measures are associated with a reduction in mortality risk over the long term: given surviving the tsunami itself, those from exposed communities are more likely to survive the 15 years

Table 1.2 Community exposure, individual exposure, and risk of mortality 2005–2020

		(1)	(2)	(3)	(4)
Tsunami exposures	Community affected	−2.91			
		[1.02]**			
	% of community killed		−0.17	−0.23	
			[0.04]**	[0.05]**	
	Direct exposure to waves			0.01	−0.32
				[1.68]	[2.04]
	Spouse killed			3.71	3.73
				[2.94]	[3.20]
	Parent, child, or sibling killed			−1.53	−1.22
				[1.31]	[1.40]
	Lived in temporary housing			5.11	3.17
				[1.45]**	[2.14]
	High level of post-traumatic stress			−0.61	−0.86
				[1.34]	[1.50]
Baseline characteristics	Male	10.87	11.17	11.01	11.38
		[1.42]**	[1.43]**	[1.42]**	[1.45]**
	Widowed or divorced	−0.53	−0.43	−0.03	−0.32
		[1.67]	[1.66]	[1.67]	[1.74]
	Education: 4–6 years	−0.56	−0.53	−0.59	−0.89
		[1.39]	[1.39]	[1.39]	[1.50]
	7–11 years	−2.12	−1.87	−2.03	−2.94
		[1.78]	[1.77]	[1.77]	[1.96]
	12+ years	−1.70	−1.88	−1.59	−2.88
		[1.63]	[1.63]	[1.62]	[1.98]
	Monthly per capita expenditure	−2.35	−2.33	−2.22	−3.14
		[0.90]**	[0.92]*	[0.91]*	[1.24]*
Constant		55.46	53.86	50.88	64.22
		[18.97]**	[19.18]**	[19.12]**	[21.98]**
Observations		5927	5927	5927	5927
R-squared		0.22	0.23	0.23	0.23
EA fixed effect		No	No	No	Yes
Number of Eas					333

Note: Robust standard errors in brackets. Controls for age (in single years)
+significant at 10%; * significant at 5%; ** significant at 1%

after the 2004 tsunami than those from communities in which impacts were minor. Age is controlled in the models (using indicators for single years), so these results do not reflect differences in age composition across communities. Nor do they reflect differences in gender, widowhood, education, or pre-tsunami resources, all of which are taken into account in the models.

Because the continuous measure of community mortality provides a more fine-grained measure of exposure relative to the dichotomous indicator, subsequent regressions control for the community tsunami mortality rate.

Moving to column 3, we add controls for individual exposures. This addition only strengthens the negative effect of the community mortality rate, which increases in magnitude by roughly 40%, from −0.17 to −0.23. Two of the individual exposure measures exhibit relatively strong positive effects on mortality: death of a spouse (3.71, albeit not statistically significant) and having lived in temporary housing (5.11, statistically significant). These results suggest that although positive mortality selection operates in the aggregate, individuals with certain experiences carry scars that affect their long-term survival prospects. The coefficients on the other exposure measures are much smaller in magnitude, and none is statistically significant.

Column 4 presents the results based on Model 3, which includes community fixed effects and thus draws comparisons between individuals who were living in the same community at the time of the tsunami. The size of the coefficient associated with living in poor housing is reduced and less precisely estimated ($p = 0.14$). This reflects the fact that there is little variation in the extent of physical damage to property within a community, relative to the variation across communities. As a result, the need for temporary housing is highly correlated across individuals who were living in the same community at the time of the tsunami.

With respect to the baseline characteristics, males have markedly higher mortality than females—by ten to eleven percentage points. Additionally, as baseline level of economic resources rises, mortality risk falls, suggesting that pre-disaster SES exerts a protective effect on health long after the event. Neither educational level nor baseline marital status is related to post-tsunami survival.

As shown in Fig. 1.1, mortality risks in the tsunami vary by age and sex. Accordingly, we estimate Models 1 and 3 separately for each of the four age-sex groups distinguished in Fig. 1.1. These results are presented in Table 1.3.

Perhaps the most important overall result from the stratified models is the substantial difference in our ability to predict post-tsunami mortality for older versus younger respondents. For younger respondents, neither the community-level mortality rate nor the measures of individual exposures are statistically significant predictors of mortality after the disaster. Nor for the most part are the measures of demographic and socioeconomic background (the one exception is per capita expenditure levels for younger males in the fixed effect specification). Post-tsunami mortality risks do rise with age for both males and females (results not shown).

For older individuals the story is markedly different. The force of positive mortality selection is very powerful for both males and females. For each 1% increase in the percentage of residents killed in the tsunami, the risk of mortality in the next 15 years falls by 0.36 percentage points for males and by 0.38 percentage points for females.

In addition, a number of the individual exposure measures elevate mortality risks of older adults. For older males three exposure measures are large in size and marginally significant. Mortality risks are 11.6 percentage points higher for men who lived in temporary housing ($p = 0.054$) and 9 percentage points higher ($p = 0.07$) for men who experienced high levels of post-traumatic stress in the first 2 years after

Table 1.3 Community exposure, individual exposure, and risk of mortality 2005–2020, by age and sex

		Males 35–49		Females 35–49		Males 50+		Females 50+	
		(1)	(2)	(3)	(4)	(5)	(6)	(7)	(8)
Tsunami exposures	% of Community killed	−0.08		0.01		−0.36		−0.38	
		[0.05]		[0.07]		[0.11]**		[0.16]*	
	Direct exposure to waves		−0.50		−1.25		−2.50		−6.91
			[3.16]		[3.13]		[6.13]		[6.66]
	Spouse killed		−5.12		6.06		−13.39		17.82
			[5.98]		[5.65]		[7.23]+		[7.92]*
	Parent, child, or sibling killed		0.86		−2.87		−4.27		5.12
			[2.82]		[2.00]		[4.29]		[4.11]
	Lived in temporary housing		0.05		2.57		11.58		7.33
			[3.53]		[4.02]		[6.02]+		[6.96]
	High level of post-traumatic stress		−2.00		−3.37		8.86		−2.23
			[2.59]		[2.13]		[4.90]+		[4.44]
Baseline characteristics	Widowed or divorced	−7.12	−12.92	0.91	3.69	−2.28	4.72	4.36	6.10
		[6.98]	[9.23]	[2.26]	[2.38]	[6.18]	[6.54]	[2.88]	[3.45]+
	Education: 4–6 years	0.69	0.61	2.64	2.64	−2.79	−4.57	−3.68	−4.27
		[2.92]	[3.24]	[2.03]	[2.34]	[3.58]	[4.19]	[2.95]	[3.65]
	7–11 years	−1.77	−2.35	1.31	1.70	−7.72	−7.44	1.48	3.40
		[3.28]	[3.88]	[2.38]	[2.83]	[4.65]+	[6.14]	[4.77]	[6.20]
	12+ years	−1.42	−2.06	1.64	2.17	−5.31	−4.98	−7.33	−13.65
		[3.00]	[3.89]	[2.19]	[2.90]	[4.72]	[5.73]	[4.83]	[6.45]*
	Monthly per capita expenditure	−2.64	−5.60	0.10	−2.57	−1.06	−1.50	−6.38	−5.54
		[1.68]	[2.48]*	[1.21]	[1.87]	[2.45]	[3.73]	[2.29]**	[3.95]

		Males 35–49		Females 35–49		Males 50+		Females 50+	
		(1)	(2)	(3)	(4)	(5)	(6)	(7)	(8)
Constant		40.64	68.84	40.37	70.81	101.26	70.48	133.18	131.41
		[30.03]	[42.33]	[25.46]	[35.02]*	[52.42]+	[69.27]	[51.07]**	[71.29]+
Observations		1799	1799	1686	1686	1242	1242	1200	1200
R-squared		0.05	0.06	0.04	0.04	0.16	0.18	0.21	0.22
EA fixed effect		No	Yes	No	Yes	No	Yes	No	Yes
Number of Eas			330		316		319		302

Note: Robust standard errors in brackets. Controls for age (in single years)
+ significant at 10%; * significant at 5%; ** significant at 1

the tsunami. On the other hand, the mortality risks of men who lost a spouse are reduced by 13.4 percentage points (p = 0.065).

Turning to older women, the loss of a spouse has exactly the opposite relationship with mortality risks. For women, losing a spouse increases the risk of subsequent mortality by almost 18 percentage points. The difference between the effect for males and females is large and statistically significant (results not shown). One possible reason for this difference is the difference in opportunities available to older males and females who are widowed. In particular, they face very different marriage markets. For the younger respondents, death of a spouse is not statistically related to mortality for either males or females, but we note that the coefficients are relatively large and the signs are the same as for older adults: negative for men but positive for women.

For older women the other exposure measures are not statistically significant, although the impact of living in poor housing is large and positive (and marginally significant in the model without the community-level fixed effect, which is not shown).

Among older respondents, background variables matter little for males, although 7–11 years of schooling reduces mortality risks relative to men with 0–4 years of education (the omitted group). For females, higher levels of economic resources before the tsunami are associated with lower levels of mortality post-tsunami, but being widowed or divorced at baseline increases mortality risks.

Post-traumatic Stress

Over and above survival, it is important to examine other dimensions of health which are likely to be indicative of quality of life and subsequent mortality. We turn, therefore, to an indicator of psychosocial well-being that is particularly salient in this context. Our earlier work documents the strong role that exposure played with respect to levels of post-traumatic stress in the first few years after the tsunami (Frankenberg et al., 2008). Here we examine the degree to which exposure continues to play a role in post-traumatic stress reactivity at 5, 10, and 15 years after the disaster.

Table 1.4 presents results from Model 2 (where the outcome variable is the index of PTSR that varies from 0 to 21) for the three time periods, estimated separately for the four age-sex groups. For each group some aspects of exposures matter, although what is important varies by time period and by group. Moreover, the exposure measures are more closely correlated with PTSR for the younger age groups than they are for the older ones (the reverse of what we see for mortality). Among males aged 35–49 years, losing a spouse increases the PTSR level at the 5 years post-tsunami mark. At 10 years, both direct exposure to the waves and being from a community with a higher mortality rate are associated with higher levels of PTSR. At 15 years, only the community mortality rate emerges as important, with a higher mortality rate associated with more symptoms.

Table 1.4 Community exposure, individual exposure, and levels of post-traumatic stress 5, 10, and 15 years after the tsunami

# of years	Males age 35–49			Females age 35–49			Males age 50+			Females age 50+		
post-tsunami	+5	+10	+15	+5	+10	+15	+5	+10	+15	+5	+10	+15
Tsunami exposures												
% of community killed	0.00	0.02	0.02	0.01	0.01	0.00	0.02	0.00	−0.01	0.03	0.05	0.05
	[0.01]	[0.01]**	[0.01]*	[0.01]	[0.01]	[0.01]	[0.01]	[0.02]	[0.01]	[0.02]	[0.02]*	[0.03]+
Direct exposure to waves	−0.14	0.63	0.15	0.09	0.14	0.61	0.45	1.39	0.56	0.38	0.86	1.39
	[0.27]	[0.24]**	[0.23]	[0.40]	[0.31]	[0.32]+	[0.48]	[0.54]*	[0.42]	[0.50]	[0.63]	[0.85]
Spouse killed	1.22	0.48	−0.59	1.28	1.86	0.44	−1.61	0.69	−0.53	−0.66	−0.12	1.23
	[0.66]+	[0.42]	[0.48]	[0.64]*	[0.72]*	[0.54]	[0.68]*	[1.02]	[0.49]	[0.65]	[0.70]	[1.02]
Parent, child, sibling killed	0.09	−0.28	−0.03	0.73	0.54	0.12	−0.38	0.73	0.32	−0.54	0.35	−0.14
	[0.28]	[0.21]	[0.20]	[0.31]*	[0.28]+	[0.20]	[0.40]	[0.37]*	[0.34]	[0.40]	[0.45]	[0.39]
Lived in temporary housing	0.39	−0.42	0.23	0.80	−0.45	0.49	−0.25	−0.18	−0.10	0.26	−0.55	0.26
	[0.27]	[0.19]*	[0.20]	[0.31]**	[0.26]+	[0.26]+	[0.40]	[0.38]	[0.39]	[0.49]	[0.45]	[0.54]
Baseline characteristics												
Widowed or divorced	−0.89	0.15	−0.04	0.12	0.14	−0.14	−1.17	0.45	−0.83	−0.28	0.12	0.00
	[0.74]	[0.96]	[0.53]	[0.31]	[0.29]	[0.21]	[0.59]*	[1.00]	[0.59]	[0.33]	[0.31]	[0.28]
Education: 4–6 years	−0.40	0.20	−0.48	0.78	−0.28	−0.03	−0.52	0.34	0.05	0.06	−0.26	0.08
	[0.31]	[0.22]	[0.22]*	[0.28]**	[0.24]	[0.20]	[0.36]	[0.32]	[0.37]	[0.35]	[0.37]	[0.34]
7–11 years	−0.95	0.38	−0.29	0.56	−0.30	−0.33	−1.95	0.20	−0.14	0.13	0.56	−0.39
	[0.35]**	[0.27]	[0.24]	[0.35]	[0.29]	[0.23]	[0.44]**	[0.43]	[0.37]	[0.59]	[0.60]	[0.45]

(continued)

Table 1.4 (continued)

# of years	Males age 35–49			Females age 35–49			Males age 50+			Females age 50+		
post-tsunami	+5	+10	+15	+5	+10	+15	+5	+10	+15	+5	+10	+15
12+ years	−1.13	0.08	−0.32	−0.13	−0.26	−0.19	−1.68	0.08	−0.25	−0.20	−0.81	−0.85
	[0.32]**	[0.22]	[0.23]	[0.30]	[0.31]	[0.25]	[0.46]**	[0.40]	[0.39]	[0.66]	[0.55]	[0.56]
Monthly per capita spending	0.46	−0.05	0.04	0.15	0.01	0.43	−0.06	0.00	0.31	0.47	−0.33	0.35
	[0.19]*	[0.16]	[0.14]	[0.19]	[0.17]	[0.14]**	[0.24]	[0.20]	[0.24]	[0.30]	[0.26]	[0.24]
Constant	1.61	5.09	2.54	6.16	5.54	−4.93	14.86	3.91	−4.27	0.16	7.42	1.82
	[3.62]	[2.74]+	[2.87]	[3.54]+	[3.10]+	[2.50]*	[5.51]**	[4.45]	[6.16]	[5.86]	[5.06]	[4.50]
Observations	1333	1407	1292	1421	1438	1382	554	541	471	652	623	569
R-squared	0.03	0.05	0.03	0.06	0.03	0.04	0.15	0.12	0.09	0.07	0.07	0.11

Note: OLS regressions, robust standard errors in brackets. Restricted to survivors to the most recent STAR follow-up, controls for age (single years)
+ significant at 10%; * significant at 5%; ** significant at 1%

Among women aged 35–49 years, the community mortality rate is not associated with PTSR levels at any point, but each of the other exposure measures affects PTSR levels in at least one period. In fact, the individual exposure measures are more closely associated with PTSR for women aged 35–49 years than for any other group. In terms of magnitude, the largest effects are for losing a spouse, which elevates the PTSR level at both 5 and 10 years after the disaster. Losing another close family member also elevates PTSR at 5 and at 10 years. Living in temporary housing is associated with higher levels of PTSR at years 5 and 15. Finally, direct exposure to the waves elevates PTSR at 15 years post-disaster. Although exposure does not translate into reduced longevity for the younger women, each of the individual-specific indicators of tsunami-related trauma continues to take a toll on this indicator of psychosocial health more than a decade later.

Among older respondents, some exposures are linked to PTSR levels, but the effects are far more muted. For older men, losing a spouse reduces PTSR symptoms 5 years after the tsunami (which matches the negative relationship that losing of a spouse exhibits with mortality), but there is no relationship between spousal loss and PTSR at other points. At 10 years, direct exposure to the waves and loss of a parent, child, or sibling increase PTSR levels. For older men, none of the exposure measures are relevant at 15 years post-tsunami. For older women, there are few correlations between exposure and PTSR. For this group, the community mortality rate is positively related to PTSR levels 10 and 15 years after the tsunami, but this is the only measure of exposure that is correlated with post-traumatic stress.

Background characteristics appear to have little impact on PTSR for any group, although for males in both age groups, higher levels of education are associated with lower levels of PTSR at 5 and 10 years after the disaster.

Table 1.5 presents the results for PTSR with the addition of a community fixed effect (Model 3). The result is a reduction in the importance of the exposure measures for all groups, particularly 10 and 15 years after the tsunami. For younger males, having lived in temporary housing affects PTSR at 5 years, but no other exposure measures are related to PTSR. Among younger females, loss of a spouse or loss of other close kin affects PTSR at years 5 and 10, but none of the exposure measures matter at year 15. For older males, direct exposure elevates PTSR at year 10, while for older females, it elevates PTSR at year 15. Thus, it appears that within communities, variation in levels of PTSR many years after the tsunami is not primarily explained by exposures at the time of the event.

Impacts on Health and Well-Being

Taken together, the results for mortality and PTSR establish that a large-scale natural disaster exerts enduring impacts on health and well-being. In communities that were directly affected by the tsunami, survivors are positively selected with respect to characteristics associated with longevity. Nonetheless, the effect of their experiences at the time of the tsunami and in the first years after the tsunami is evident

Table 1.5 Individual exposure and levels of post-traumatic stress after the tsunami with enumeration area fixed effects

# of years	Males age 35–49			Females age 35–49			Males age 50+			Females age 50+		
post-tsunami	+5	+10	+15	+5	+10	+15	+5	+10	+15	+5	+10	+15
Tsunami exposures												
Direct exposure to waves	−0.28	0.41	−0.11	0.49	−0.07	0.60	−0.09	1.58	0.70	0.64	−0.05	2.37
	[0.31]	[0.25]	[0.26]	[0.44]	[0.36]	[0.42]	[0.61]	[0.79]*	[0.93]	[0.66]	[0.84]	[1.03]*
Spouse killed	1.32	0.70	−0.06	0.47	2.28	0.68	−1.15	−0.10	−0.64	−0.78	0.02	−0.85
	[0.81]	[0.46]	[0.60]	[0.72]	[0.78]**	[0.70]	[0.87]	[1.24]	[0.73]	[0.86]	[0.97]	[0.91]
Parent, child, or sibling killed	0.26	−0.37	−0.07	0.63	0.53	0.00	0.48	0.82	0.09	−0.26	0.71	−0.1
	[0.33]	[0.25]	[0.20]	[0.31]*	[0.29]+	[0.23]	[0.51]	[0.51]	[0.46]	[0.54]	[0.56]	[0.55]
Lived in temporary housing	0.93	−0.08	0.00	0.48	−0.14	−0.28	0.91	0.67	0.58	0.11	−0.09	1.83
	[0.44]*	[0.37]	[0.29]	[0.53]	[0.42]	[0.40]	[0.79]	[0.96]	[0.60]	[0.87]	[0.75]	[1.53]
Baseline characteristics												
Widowed or divorced	−1.02	0.37	0.04	0.28	0.23	−0.09	−0.32	1.21	0.01	0.04	0.16	−0.13
	[0.87]	[1.08]	[0.61]	[0.30]	[0.29]	[0.24]	[0.82]	[1.44]	[0.58]	[0.36]	[0.42]	[0.35]
Education: 4–6 years	−0.49	0.09	−0.20	0.75	−0.40	0.04	−1.08	0.14	0.49	0.09	−0.24	0.32
	[0.33]	[0.23]	[0.21]	[0.29]*	[0.26]	[0.22]	[0.42]*	[0.48]	[0.39]	[0.38]	[0.42]	[0.43]
7–11 years	−0.58	0.37	−0.15	1.14	−0.55	−0.28	−2.25	0.20	0.06	0.22	1.14	−0.52
	[0.39]	[0.28]	[0.25]	[0.35]**	[0.32]+	[0.24]	[0.51]**	[0.66]	[0.50]	[0.67]	[0.70]	[0.68]
12+ years	−0.63	−0.12	−0.24	0.85	−0.53	−0.01	−1.01	−0.22	−0.38	0.48	−0.47	−0.36
	[0.37]+	[0.25]	[0.24]	[0.36]*	[0.33]	[0.27]	[0.65]	[0.70]	[0.64]	[0.82]	[0.66]	[0.82]
Monthly per capita expenditure	0.36	−0.06	−0.15	−0.23	0.13	0.05	0.75	0.29	−0.46	0.78	0.12	0.57
	[0.25]	[0.20]	[0.16]	[0.28]	[0.25]	[0.19]	[0.43]+	[0.32]	[0.41]	[0.43]+	[0.40]	[0.39]
Constant	0.09	4.29	2.85	3.45	2.58	−0.34	−1.44	1.27	2.11	−1.81	6.85	−9.11
	[4.32]	[3.35]	[2.97]	[4.65]	[4.38]	[2.94]	[7.31]	[7.13]	[6.68]	[7.51]	[7.23]	[7.52]
Observations	1333	1407	1292	1421	1438	1382	554	541	471	652	623	569
R-squared	0.03	0.03	0.01	0.03	0.04	0.02	0.16	0.18	0.15	0.15	0.11	0.14

Note: EA-level fixed effects. Robust standard errors in brackets. Restricted to survivors to the most recent STAR follow-up. Controls for age (single years)
+ significant at 10%; * significant at 5%; ** significant at 1%

over the long term, both in terms of their survival rates and their psychosocial health. Several findings suggest important questions for future research.

First, the differences between older males and females in the impact of losing a spouse are striking and warrant additional research on differences by sex in how lives unfold after a disaster. It is not clear how much of these differences can be attributed to differences in remarriage after loss of a spouse in the tsunami, relationships with children and other family members, evolution of economic circumstances, or the availability of assistance from family, the community, or the government. An important policy issue that has received little attention in the literature on disaster relief revolves around the design of assistance and support programs that mitigate the large negative consequences of premature spousal death for women.

Second, exposure to poor housing conditions is indicated as a risk factor for mortality and poor psychosocial health for males and females across the entire adult age spectrum. This points to the potential of well-designed housing assistance programs that are deployed soon after a destructive natural disaster to substantially reduce the negative health consequences of the disaster.

Third, there are long-lasting impacts of exposure to disaster-related trauma on psychosocial health. Even 15 years after the tsunami event, PTSR is shaped by what happened during and after the disaster. The continued role these exposures play is strongest for younger women, for whom loss of either a spouse or another close family member negatively affects psychosocial health, but it is present to varying degrees for all demographic groups. Access to mental health services was extremely limited after the tsunami. The literature suggests that deploying services on a broader scale soon after the disaster limits some of the long-term repercussions of the experience.

We close with two observations. First our data and methods provide unusual and important evidence on the causal impact of exposure. It is, however, of substantial interest—and of great importance—to investigate the mechanisms underlying heterogeneity in these outcomes and isolate the factors that seem to be associated with greater resilience and recovery in the aftermath of the tsunami.

Second, although high mortality disasters are relatively rare in high-income countries, extreme events are on the rise worldwide. The United States has seen numerous hurricanes and wildfires in recent years—events that generate the kinds of exposures we consider, even if mortality associated with them is low. Events that cost lives, damage property, and expose people to potentially traumatic experiences will punctuate life for the foreseeable future, and it is important to study their short- and long-term implications for health and well-being.

Acknowledgments The project has been supported by the National Institute for Child Health and Human Development (R01HD052762, R01HD051970, R03HD071131, P2C HD050924), the National Institute on Aging (R01AG031266), the National Science Foundation (CMS-0527763), the Hewlett Foundation, the World Bank, the MacArthur Foundation (05-85158-000), and the Wellcome Trust (OPOH 106853/A/15/Z).

References

Borrero, J. (2005). Field data and satellite imagery of tsunami effects in Banda Aceh. *Science, 308*(5728), 1596. https://doi.org/10.1126/science.1110957

Dalgleish, T. (1999). Cognitive theories of posttraumatic stress disorder. In W. Yule (Ed.), *Post-traumatic stress disorders: Concepts and therapy* (pp. 193–222). Wiley.

Deaton, A. (1997). *The analysis of household surveys: A microeconomic approach to development policy*. Johns Hopkins University Press.

Frankenberg, E., Friedman, J., Gillespie, T., Ingwersen, N., Pynoos, R., Rifai, I. U., Sikoki, B., Steinberg, A., Sumantri, C., Suriastini, W., & Thomas, D. (2008). Mental health in Sumatra after the tsunami. *American Journal of Public Health, 98*(9), 1671–1677. https://doi.org/10.2105/ajph.2007.120915

Frankenberg, E., Gillespie, T., Preston, S., Sikoki, B., & Thomas, D. (2011). Mortality, the family and the Indian Ocean tsunami. *Economic Journal, 121*(554), F162–F182. https://doi.org/10.1111/j.1468-0297.2011.02446.x

Frankenberg, E., Sumantri, C., & Thomas, D. (2020). Effects of a natural disaster on mortality risks over the longer-term. *Nature Sustainability, 3*, 614–619. https://doi.org/10.1038/s41893-020-0536-3

Ho, J. Y., Frankenberg, E., Sumantri, C., & Thomas, D. (2017). Adult mortality five years after a natural disaster. *Population Development Review, 43*(3), 467–490. https://doi.org/10.1111/padr.12075

Kohl, P., O'Rouke, A., Schmidman, D., Dopkin, W., & Birnbaum, M. (2005). The Sumatra-Andaman earthquake and tsunami of 2004: The hazards, events, and damage. *Prehospital and Disaster Medicine, 20*(6), 355–363.

Lavigne, F., Paris, R., Grancher, D., Wassmer, P., Brunstein, D., Vautier, F., Leone, F., Flohic, F., De Coster, B., Gunawan, T., Gomez, C., Setiawan, A., Cahyadi, R., & Fachrizal. (2009). Reconstruction of tsunami inland propagation on December 26, 2004 in Banda Aceh, Indonesia, through field investigations. *Pure and Applied Geophysics, 166*(1–2), 259–281. https://doi.org/10.1007/s00024-008-0431-8

Monecke, K., Finger, W., Klarer, D., Kongko, W., McAdoo, B. G., Moore, A. L., & Sudrajat, S. U. (2008). A 1,000-year sediment record of tsunami recurrence in northern Sumatra. *Nature, 453*(7217), 1232–1234. https://doi.org/10.1038/nature07374

Ramakrishnan, D., Ghosh, S. K., Raja, V. K. M., Chandran, V., & Jeyram, A. (2005). Trails of the killer tsunami: A preliminary assessment using satellite remote sensing technique. *Current Science, 88*(5), 709–711.

Rubin, C. M., Horton, B. P., Sieh, K., Pilarczyk, J. E., Daly, P., Ismail, N., & Parnell, A. C. (2017). Highly variable recurrence of tsunamis in the 7,400 years before the 2004 Indian Ocean tsunami. *Nature Communications, 8*, 16019. https://doi.org/10.1038/ncomms16019

Thiery, W., Lange, J., Rogelj, S., Schleussner, C.-F., Gudmundsson, S. I., Seneviratne, M., Andrijevic, K., Frieler, K., Emanuel, T., Geiger, D. N., Bresch, F., Zhao, S. N., Willner, M., Büchner, J., Volkholz, N., Bauer, J., Chang, P., Ciais, M., Dury, L., et al. (2021). Intergenerational inequities in exposure to climate extremes. *Science, 374*(6574), 158–160. https://doi.org/10.1126/science.abi7339

Umitsu, M., Tanavud, C., & Patanakanog, B. (2007). Effects of landforms on tsunami flow in the plains of Banda Aceh, Indonesia, and Nam Khem, Thailand. *Marine Geology, 242*, 141–153. https://doi.org/10.1016/j.margeo.2006.10.030

Weathers, F. W., Litz, B. T., Herman, D. S., Huska, J. A., & Keane, T. M. (1993). *The PTSD checklist (PCL): Reliability, validity, and diagnostic utility* [poster presentation]. Annual meeting of the International Society for Traumatic Stress Studies. https://www.researchgate.net/publication/291448760_The_PTSD_Checklist_PCL_Reliability_validity_and_diagnostic_utility.

Chapter 2
Evidence-Based Interventions for Children and Families During Disaster Recovery: Trends, Lessons Learned, and Future Directions

Tara Powell, Jenna M. Muller, and Greta Wetzel

Background

In the past 20 years, climate-induced disasters have significantly increased in severity and frequency, adversely impacting the global population. Children and adolescents represent approximately 175 million of those affected by environmental disasters annually (Dyregrov et al., 2018), and it is estimated that 11% of young people have experienced a disaster before the age of 16 (Cadamuro et al., 2021).

Children and youth are among the most vulnerable during and after a disaster due to dependence on parents and caregivers to meet their physical and emotional needs (Felix et al., 2020). In addition to disruption of everyday life, disasters often result in secondary stressors such as family conflict and/or violence, school closures, illness, injury, or parental divorce (Peek, 2008). Other disaster-related adversities include family separation, displacement from their home and community, disrupted social supports, loss of loved ones, and lack of access to basic needs including food, water, and housing (Becker-Blease et al., 2010; Kousky, 2016; Wang et al., 2013). These extreme events can also have a sustained impact on children and youths' mental health. Psychological symptoms include post-traumatic stress symptoms (PTSS), depression, and anxiety and may persist well into the disaster recovery (Cadamuro et al., 2021). These psychological challenges have also been associated with impaired development, learning, and social-emotional abilities (Arshad et al., 2020).

T. Powell (✉) · J. M. Muller
School of Social Work, University of Illinois at Urbana-Champaign, Urbana, IL, USA
e-mail: tlpowell@illinois.edu; jmuller2@illinois.edu

G. Wetzel
Save the Children, Arlington, VA, USA
e-mail: gwetzel@savechildren.org

S. E. Ortiz et al. (eds.), *Environmental Impacts on Families*, National Symposium on Family Issues 12, https://doi.org/10.1007/978-3-031-22649-6_2

Families play an important role in a child's adjustment after a disaster by providing routines, a stable environment, safety, and security. Considering the critical role of the family in children and youths' post-disaster adaptation, we provide an overview of the role of the family system and social-ecological risk and protective factors that contribute to children and youths' recovery. We then review post-disaster evidence-based interventions and provide an overview of the Journey of Hope, a disaster-focused social-emotional program. We conclude this chapter discussing challenges and future directions of intervention delivery and research in post-disaster environments.

Social-Ecological Risk and Protective Factors

Post-disaster psychological adjustment is based on numerous factors that influence children and youths' ability to overcome these extreme events. Resilience, or the ability to successfully adapt to a threatening situation, plays an important role in post-disaster adaptation and is dependent on individual-, relationship-, and community-level social and ecological risk and protective factors (Masten & Barnes, 2018). Individual characteristics that increase risk for post-disaster psychopathology among children include previous trauma exposure, pre-existing mental health difficulties (e.g., depression, anxiety), female gender, younger age, and maladaptive coping strategies (Cadamuro et al., 2021). A systematic review, for example, found that avoidant and ruminative coping styles, negative affect, and pre-disaster psychopathology were all predictors of post-disaster PTSS symptomology (DiGangi et al., 2013). Conversely, individual-level protective factors that increase the likelihood of a child adapting or overcoming disaster-related adversity include positive or active coping strategies, emotional and behavioral regulation, and positive sense of self/self-esteem (DiGangi et al., 2013). One study that was conducted after a flood in Canada, for example, found that children who experienced fewer externalizing and internalizing behavior problems reported significantly higher resilience scores than those with more challenges (Arshad et al., 2020). While some individual-level risk and protective factors are unalterable such as age and gender, others such as coping, self-esteem, and emotion regulation are modifiable and influenced by social and family factors.

Relationship factors such as family characteristics and social connection also play an important role in post-disaster adaptation and influence individual-level modifiable risk and protective factors. Higher levels of PTSS, for example, have been reported among children whose parents are experiencing distress and are unable to provide emotional and instrumental support during and after a disaster (Cadamuro et al., 2016; Felix et al., 2016). A study after hurricane Katrina found children with low perceived family connectedness experienced sustained depression and PTSS symptoms (Kronenberg et al., 2010). Other modifiable parent/caregiver risk factors include family conflict, low perceived family connectedness or support, and parental psychopathology (Cobham et al., 2016; Zhou, 2018).

Parental/caregiver connectedness, a stable home environment, and social connections are all associated with reduced risk of psychopathology among disaster-affected children and youth (Arshad et al., 2020; Cicchetti, 2010; Masten, 2013). Similarly, a positive relationship between perceived social support and resilience in children who experienced a flood disaster has been shown (Arshad et al., 2020). This social support may come from different sources; peer support, for example, is an important way for children and youth to process shared experiences and re-establish stability and normalcy (McDonald-Harker et al., 2021). Studies have also illustrated that family hardiness, which includes having a sense of control after stressful situations and viewing hardships as opportunities to grow, can protect against short- and long-term post-disaster psychological distress (Hackbarth et al., 2012; Jovanovic et al., 2004). For example, family hardiness and hope have been found to increase coping ability in hurricane-affected individuals (Hackbarth et al., 2012). Other studies have indicated that parent's/caregiver's ability to stay calm, maintain a positive affect, and remain consistent with routines are all associated with better adjustment and fewer psychological symptoms in their children (Chen et al., 2020; Zhou, 2018).

Children and families' post-disaster adaptation is also influenced by the larger community systems in which they live. Lack of community cohesion, socioeconomic status, school closures, and fractured social infrastructure are all factors that increase the risk for sustained psychopathology. Those who live in poverty, for example, are often less equipped for these emergencies, are at higher risk for displacement and family separation, and often require assistance to meet their basic needs (Rufat et al., 2015; Vu & VanLandingham, 2012). During the recovery phase of a disaster, economically vulnerable populations are at a disproportionate risk of mental and physical health pathology such as contracting diseases due to poor living conditions and experience higher rates of depression and PTSS than those with greater socioeconomic means (Amaratunga & O'Sullivan, 2006; Boscarino et al., 2014; Gutman & Yon, 2014; Jia et al., 2010; Kouadio et al., 2012; La Greca et al., 2013; Parkinson & Zara, 2013). Moreover, extended school closures after a disaster can disrupt access to services and impede a young person's ability to learn and gain vital social-emotional skills (Masten, 2021).

Community-level protective factors also play a central role in post-disaster recovery. Community-based organizations, schools, and religious organizations all provide essential resources to help children and families recover from a disaster (Masten, 2021). Schools, for example, provide critical health and psychological supports and a stable environment while re-establishing a sense of safety and security among children and families in post-disaster contexts (Mooney et al., 2020). See Fig. 2.1 for a visualization of social-ecological risk and protective factors.

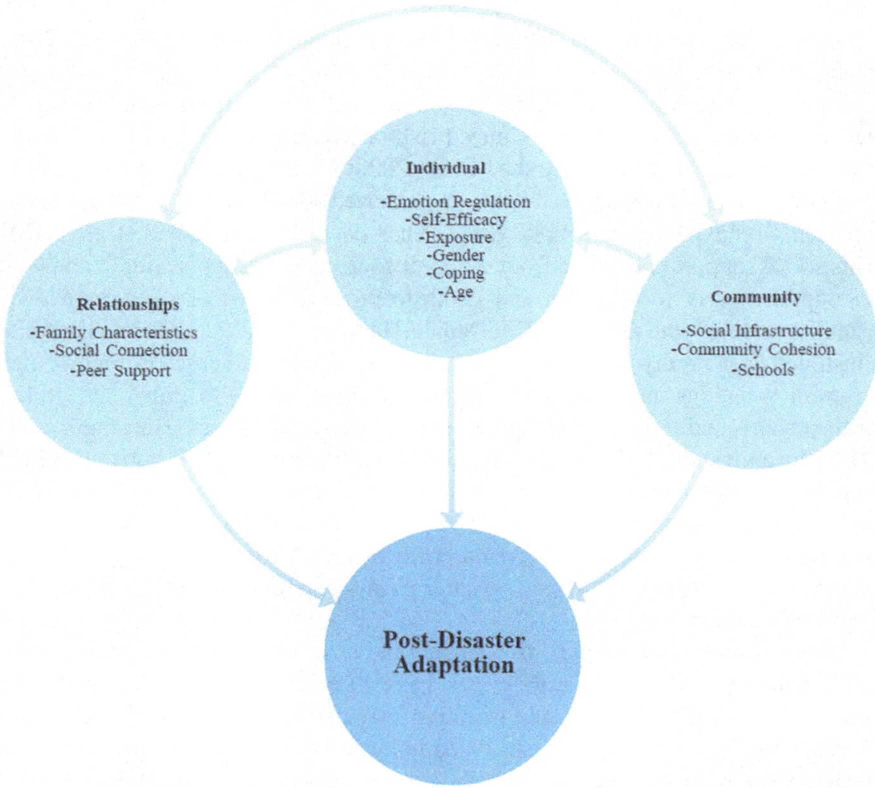

Fig. 2.1 Social-ecological interaction of risk and protective factors

Evidence-Based Interventions in Post-Disaster Contexts

Considering the impact of disasters on children and families and the role of risk and protective factors in post-disaster adaptation, evidence-based interventions have been developed to treat and/or reduce the risk of short- and long-term psychopathology. These interventions are designed to be delivered immediately after a disaster or during the longer-term recovery period.

In the immediate aftermath of a disaster, psychological first aid (PFA) is a commonly used evidence-informed approach to support the psychological needs of children and families. While not a treatment modality, PFA is often employed by first responders to reduce initial psychological distress among survivors. Five components of PFA include (1) ensuring survivors are safe; (2) providing comfort; (3) connecting individuals to essential resources; (4) promoting self-efficacy; and (5) encouraging hope (Shultz & Forbes, 2014). While PFA can be provided to adults and youth, developmental considerations such as language comprehension and reading ability are essential when delivering the intervention to children and adolescents (Gilbert et al., 2021).

Most treatment interventions provided during the recovery period target post-traumatic stress symptoms and include the use of cognitive behavioral approaches (CBT), eye movement desensitization and reprocessing (EMDR), and narrative exposure therapy (Brown et al., 2017). These evidence-based interventions are delivered in individual or group-based settings and generally include multiple sessions to treat symptoms. Trauma-focused cognitive behavioral therapy (TF-CBT), for example, is a treatment modality that incorporates psychological, cognitive, and behavioral techniques to help children regulate emotions. Parental psychoeducation is included in TF-CBT to increase family engagement, communication, and parenting skills (Cohen & Mannarino, 2008). Additionally, TF-CBT incorporates a trauma narrative that can be done through journaling, play, or arts to enable the child to tell the story of their experience (Cohen & Mannarino, 2008).

A second evidence-based treatment modality, EMDR, incorporates both cognitive and exposure components (Lewey et al., 2018). Relaxation and visualization exercises guide individuals through distressing trauma-related thoughts and replace them with positive cognitions. A primary feature of EMDR is having a client identify a "safe place" while encouraging them to continue thinking of a negative feeling or emotion associated with a trauma. The use of EMDR has been found effective in helping children identify and distinguish between positive and negative cognitions related to a potentially distressing event (Ahmad & Sundelin-Wahlsten, 2008). Narrative exposure therapy also has a well-established evidence base for both child and adult disaster survivors who are experiencing PTSS (Robjant & Fazel, 2010). The child version of narrative exposure therapy, KIDNET, incorporates child-appropriate cognitive behavior techniques in addition to writing a chronological narrative of the traumatic event (Brown et al., 2017; Fazel et al., 2020).

Prevention interventions have also been developed to mitigate the risk of disaster-related psychological distress in children and families during the disaster recovery. These interventions, often delivered in classroom-based settings and with groups of children, generally focus on mental health promotion and problem-solving skills and address social and behavioral difficulties (Higgen et al., 2021). Schools are the primary setting for delivery of prevention programs because they provide an accessible and natural way to reach children and youth who would not otherwise receive behavioral health services (Ager et al., 2010; Wolmer et al., 2011). These interventions, which target a variety of mental health challenges, can reduce externalizing, internalizing, peer, and attention difficulties. In addition, preventative interventions may increase protective factors, such as healthy coping, peer prosocial behaviors, and emotion and behavioral regulation, serving as a mechanism through which a child's likelihood of developing clinical manifestations of anxiety, depression, and post-traumatic stress may be attenuated (Pfefferbaum et al., 2015; Sanchez et al., 2018; Weare & Nind, 2011). Most post-disaster prevention and treatment interventions are beneficial in reducing distress among children and youth. However, few address the larger social ecological context and engage the family in the post-disaster recovery process. One program does address the needs of families during the disaster recovery. Journey of Hope is a prevention intervention designed to support post-disaster psychological adjustment among parents/care providers, children, and adolescent disaster survivors.

Journey of Hope: Post-Disaster Intervention

Journey of Hope (JoH), developed after Hurricane Katrina struck the Gulf Coast of the United States in 2005, was created in response to the need for intervention models to prevent and reduce hurricane-related emotional distress among children and families. The program was conceived after a gang fight in a public school that led to the potential expulsion of 20 middle school students in 2006. To prevent the expulsion of the students, the Recovery School District in New Orleans contacted social workers from the humanitarian response organization, Save the Children (SC), to provide social and behavioral support counseling to the students involved in the altercation. As the social workers provided clinical group work services to the youth, they realized the lack of available disaster-focused social-emotional programming in the school district. Interventions were available to treat disaster-related symptomology such as PTSS, depression, and anxiety. However, there was a lack of generalized prevention interventions to address the social-emotional needs of youth during the hurricane recovery. Many of the children in post-Katrina New Orleans experienced significant emotional and behavioral regulation challenges, not only from the hurricane, but due to secondary traumas such as community violence and separation from, or loss of, family members.

To address these challenges, three SC social workers documented discussions and piloted a series of social-emotional learning activities with the youth. Additionally, they conducted psychoeducational workshops with parents and teachers to identify specific needs of the families in post-Katrina New Orleans. These clinicians worked directly with the youth and their families over the course of a year, documenting relevant and effective activities. However, they did not have expertise in curriculum design. Therefore, Save the Children established a partnership with the International Institute of Child Rights and Development to guide the formal development of the Journey of Hope (JoH) curricula. This collaborative partnership also led to the first evaluation of the JoH program in 2009. Through this research-practice partnership, the researchers and clinicians worked together over the course of a year to design and pilot the JoH curricula. The finalized JoH intervention model yielded five developmentally appropriate child-level curricula and a parent/caregiver workshop.

Journey of Hope: Intervention Description

Grounded in social cognitive and stress and coping theories (Benight & Bandura, 2004; Lazarus & Folkman, 1984), the JoH evidence-based intervention model integrates parent and child sessions using a group-based approach to increase protective factors such as social connectedness, emotional awareness and regulation, problem-solving skills, and adaptive coping (see Fig. 2.2; Bandura, 1998; Lazarus & Folkman, 1984). The child-level JoH consists of eight 1-hour sessions with groups

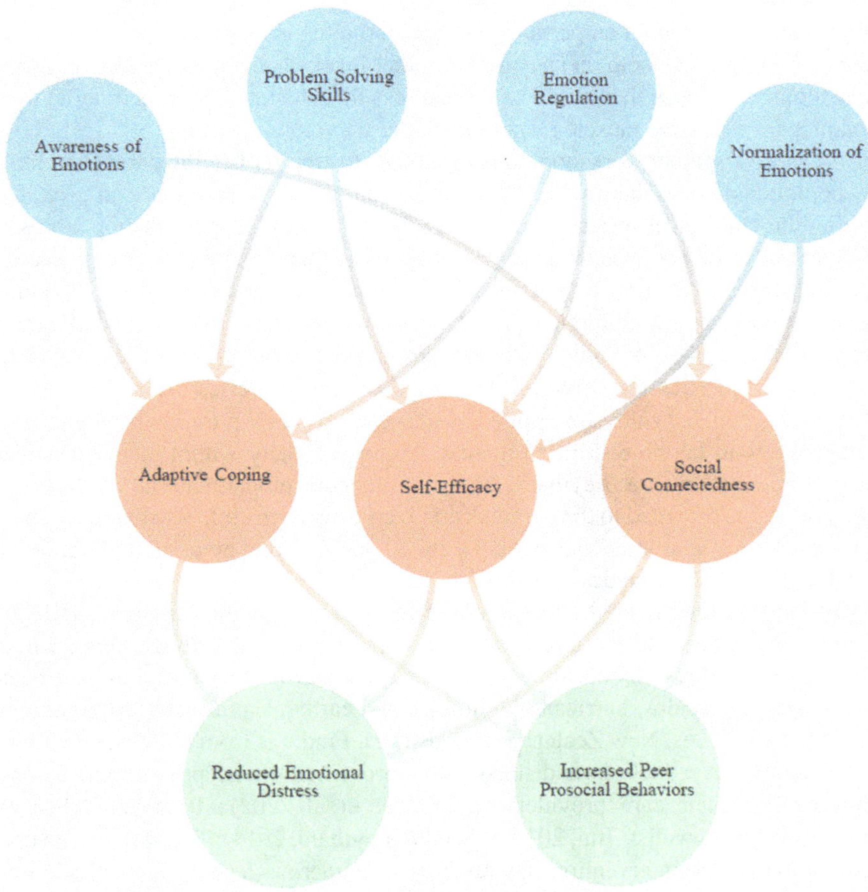

Fig. 2.2 Core principles of Journey of Hope

of 8–10 children/youth and focuses on coping with and regulating challenging emotions (e.g., fear, anger, grief) that are often heighted after a traumatic experience. Intervention delivery mechanisms include discussion, cooperative play, literacy, and mindfulness exercises. Social cognitive techniques are used to facilitate discussion of shared experiences and modeling of positive peer social interaction, thereby increasing self-efficacy, social connectedness, and adaptive coping.

The fourth session, which focuses on identifying and managing anger and aggression, for example, begins with a check-in, overview of the topic of the day, and review of group guidelines. An interactive cooperative game on anger regulation is then facilitated with the group, followed by a discussion on identifying and establishing healthy coping strategies to express anger. During the discussion, the facilitators reinforce healthy strategies to cope with anger such as deep breathing and talking to someone and strategize ways to reduce anger triggers. A second cooperative game is facilitated to employ healthy strategies that arose during the discussion. The group then participates in a journaling activity in which they write, draw, or compose a poem on an experience that invoked anger and healthy or adaptive coping strategies they could use to express their feelings. The facilitators then conduct a progressive muscle relaxation exercise to close the group.

A caregiver/parent workshop is also provided to amplify healthy parent and family coping and equip them with skills and resources to reinforce material provided during the child-level sessions. The objectives of the caregiver workshop are twofold: (1) increase parents/caregiver post-disaster adaptation and (2) increase capacity to support children's post-disaster adjustment. The workshop includes psychoeducation on children's reactions to trauma and trauma-related stress responses and provides individual- and community-level resources to cope with disaster-related stressors. Mindfulness and breathing exercises, individual reflection, and group activities are applied within each workshop to equip parents and caregivers with tangible skills to increase coping capacity within themselves and their children (Powell & Leytham, 2014). Psychoeducational material on the emotion covered is provided to the parents/caregivers after each child session (e.g., anxiety, anger) as well as strategies to help a child cope with that emotion (see Table 2.1 for detailed intervention description).

Since its development, the JoH has been adopted by Save the Children and delivered in settings around the globe in response to climate-induced disasters such as hurricanes, wildfires, floods, tornadoes, and earthquakes. Efficacy studies have been conducted in tornado-, hurricane-, wildfire-, and earthquake-affected communities in the United States, New Zealand, and Australia. Findings from these studies illustrated significant reductions in distress and improvements in adaptive coping among children and their care providers (Alexander et al., 2021; Blanchet-Cohen & Nelems, 2013; Powell & Bui, 2016; Powell & Leytham, 2014). The widespread dissemination of the intervention has resulted in numerous lessons learned such as appropriate cultural and contextual adaptations and how to maintain fidelity in ever-changing post-disaster contexts.

Scaling Up and Maintaining Fidelity

Evidence-based practice assumes that an intervention is being implemented in accordance with its published details. Therefore, oversight and support for delivery of the JoH are essential to maintain program fidelity (Carroll et al., 2007). To maintain fidelity, individuals who facilitate the intervention complete an assessment form after each session, and staff from Save the Children passively observe two of the eight sessions. The observer appraises components of program delivery including group facilitation techniques, delivery of core elements, and rapport with child participants. After the session the facilitators debrief with the observer to provide feedback on the session and discuss challenges or successes with program delivery. As part of the debriefing, the observers use a fidelity-check-guidance document to

Table 2.1 Journey of Hope description

Children and youth sessions	
Safety	Program overview, check-in, ice-breaker cooperative game, establishing group guidelines, discussion on safety, art activity-safety map, closing activity: visualizing my safe place
Fear	Check-in, introduction of topic, cooperative game, psychoeducation on fear, group discussion, cooperative game: Identifying, understanding, and coping with fear, journaling, closing mindfulness activity
Worry/anxiety	Check-in, introduction of topic, cooperative game, psychoeducation on worry/stress/anxiety, cooperative game: Trust circle, journaling, closing activity: Progressive muscle relaxation
Grief/sadness	Check-in, introduction of topic, cooperative game, psychoeducation on grief/sadness, group discussion, cooperative game: Identifying, understanding, and coping with grief or sadness, journaling, closing visualization activity
Anger	Check-in, introduction of topic, cooperative game, psychoeducation on anger and aggression, group discussion, cooperative game: Feeling and expressing anger, journaling, closing activity: Progressive muscle relaxation
Peer conflict	Check-in, introduction of topic, cooperative game, psychoeducation on peer conflict, discussion, cooperative game: Skit on managing peer conflict, journaling, closing deep breathing activity
Self-esteem	Check-in, introduction of topic, cooperative game, psychoeducation on self-esteem and self-efficacy, group discussion, cooperative game: "What you like about me," journaling, closing visualization activity
My community	Check-in, introduction of last session, closing the group, cooperative game: "Teach me," discussion: "Creating my community," journaling, closing celebration
Caregiver sessions	
Caregiver workshop	Psychoeducation: children's reactions to stress, types and sources of stress, how stress affects the body, strategies for self-care and coping, mindfulness exercises, group activity on how to collectively support each other during collective trauma recovery
Caregiver engagement	Tip sheets on session topics and 15-min discussions with parent/caregivers after each session on how to help children cope with topic (emotion) of the day

Note: Created from Save the Children (2022)

facilitate the discussion which includes (1) asking open-ended questions and allowing the facilitators to guide the conversation, (2) offering suggestions and feedback on how to address the challenges they are facing in a supportive and collaborative way, and (3) identifying any issues for follow-up.

Cultural and Contextual Adaptations to Intervention

The JoH has been delivered across the globe in New Zealand, Australia, Ukraine, Canada, the United Kingdom, and Spain. Within these countries, the program has been delivered to refugees and indigenous peoples (e.g., Maori in New Zealand;

First Nations in Canada). Thus, adaptation is imperative to ensure the program is culturally and contextually relevant for these individuals and communities.

A standardized cultural adaptation guide (see Table 2.2) was developed through a collaboration between a university researcher who was one of the original developers of the JoH and Save the Children staff in the United States, United Kingdom, New Zealand, Canada, and Australia. The cultural adaptation guide incorporates community-based participatory research principles such as involving community members in the adaptation process, employing a multi-disciplinary team to have a comprehensive set of perspectives, and including local and institutional actors as valuable contributors of knowledge and resources (Collins et al., 2018). Accounting for culture, language, and values is also a critical component of the adaptation process. The guidance document also advises that the process of adaptation must be

Table 2.2 Cultural adaptation guide

Tasks	Key considerations
Complete the community assessment tool	Assess need, language customs, and local capacity
Determine target audience	Who are the beneficiaries of your programming?
Determine available programming	What programs exist for your target audience? Are they culturally appropriate?
Determine language translation	Into what languages should the material be translated? Are there local staff you can hire to complete the translation?
Select a panel of local experts	Select local and country-specific experts. This should include a panel of experts such as: ⇒ Local government officials. ⇒ Teachers or college professors. ⇒ Professional researchers. ⇒ Community service workers. ⇒ Ministers and other religious leaders. ⇒ Parents/guardians/caregivers.
Obtain feedback from local experts	The local expert panel should review the program and associated materials for cultural appropriateness and to identify what modifications should be made to adapt the program while maintaining its fidelity. Provide reviewers with a list of considerations, including: ⇒ Is the material relevant to the needs of the beneficiaries? ⇒ Is the program transferrable—Can local staff deliver the program? ⇒ Can the text be literally translated? ⇒ Does the content specifically address the needs of the community? ⇒ How can the program be culturally modified to fit the needs of the community? ⇒ Can activities be adapted to promote cultural sensitivity? ⇒ Is additional evaluation necessary to assess whether the program maintains effectiveness with the population?.
Identify mode of delivery	What is the most appropriate mode of delivery? (e.g., in schools, community centers, displacement centers)

Note: Created from Save the Children (2014)

collaborative, organized, planned, and iterative, with participation of those from the population for whom the adaptation is targeted.

The Journey of Hope was designed with specific methodology, theory, and practice elements. Therefore, the cultural adaptation process provides guidance on maintaining core elements of the program while also addressing unique cultural and contextual considerations. Three steps are included in the cultural adaptation of the JoH: (1) conduct a situation assessment to determine applicability of the program in context, (2) consult local experts and community stakeholders on applicability of the program in the community, and (3) address language and cultural values of the local context (Save the Children, 2014).

Cultural Adaptation Case Example: New Zealand On February 22, 2011, a 6.3 magnitude earthquake struck Christchurch, New Zealand. The earthquake was declared the deadliest to strike New Zealand in 80 years, triggered widespread damage, displaced thousands, and caused extensive loss of life (Ardagh et al., 2012). As aftershocks continued to strike the city, many children and families experienced difficulties coping with the ongoing disruption and uncertainty. In response to the distress of the community, Save the Children New Zealand partnered with Save the Children US to adapt and deliver the Journey of Hope program to children and families affected by the earthquake.

Using the cultural adaptation guidance document as a resource, the Dynamic Adaptation Process (DAP; Aarons et al., 2012) was employed to fit the culture and context of post-earthquake New Zealand. The DAP is an approach that recognizes the needs for adaptations and a process that enables intervention adaptations while maintaining fidelity of the core elements. The DAP adaptation process included four phases: (1) exploration/assessing stakeholder characteristics and reviewing materials identifying core elements of the program, (2) preparation/adaptation of the curricula, (3) implementation/piloting the adapted program, obtaining feedback on the adapted intervention, and (4) sustainment/conducting ongoing assessments of facilitator feedback and gathering participant satisfaction surveys (Aarons et al., 2012).

The exploration and preparation phases included recruiting a panel of mental health professionals to review the curricula and provide insight on the specific needs of earthquake-affected children and families. This consultation consisted of a committee of key local stakeholders who were all New Zealand nationals, including staff from Save the Children New Zealand, psychologists, and social workers to ensure that the curriculum was adapted to the Christchurch context while maintaining fidelity to the core elements of the curricula.

The panel suggested specific alterations to fit the cultural context of Christchurch including (1) adapting the literacy activities including books that are used in the child-level curricula, (2) changing the language of the curricula from US English to New Zealand English, and (3) changing the titles to appropriately reflect the terminology of the education system (i.e., "Elementary Journey of Hope" to "Primary Journey of Hope").

The implementation phase included dissemination of the program to children and caregivers in ten earthquake-affected schools and community centers and a pilot study to examine the efficacy and acceptability of the adapted program. The pilot study used a pre- and post-test design and included N = 106 parents and N = 184 children and youth who took part in the program. Findings illustrated that children and youth who participated in the program experienced significant reductions in emotional distress, and parents and caregivers reported increased knowledge about healthy coping strategies, community support, and stress management (see Powell & Leytham, 2014 for full study details). The sustainment phase included ongoing dissemination, ad hoc adaptations, and assessment. This phase continued until 2018 when Save the Children concluded earthquake recovery efforts in Christchurch.

Challenges to Research and Services in Post-Disaster Contexts

Many obstacles exist for researchers and clinicians in post-disaster contexts that inhibit high-quality research and service provision. Disasters are often unexpected, making data collection prior to the event unfeasible (Gilbert et al., 2021). There are also ethical concerns to conducting research in emergency contexts given the vulnerability of survivors. Quality data collection is often stymied due to recruitment challenges, high participant attrition, and difficulty in obtaining funding rapidly (Grolnick et al., 2018; Powell et al., 2021). Gathering evidence for PFA, for example, has been notoriously challenging due to complexities in the design and dissemination of the intervention approach (Shultz & Forbes, 2014). This is in part due to the chaotic and unstable environments in emergency settings and informal delivery structure of PFA. Intervention research during the disaster recovery period is also met with inherent challenges. Conducting randomized control trials (RCTs) is often unfeasible because all individuals were exposed to the trauma, and withholding an intervention is considered unethical. Additionally, those who are most affected often reside in temporary housing and are highly mobile, which impedes researchers from collecting follow-up data. Service provision in post-disaster contexts carries similar challenges, including lack of immediate or long-term funding and high participant attrition. Funding challenges in particular can serve as a barrier to recovery and may slow or completely immobilize certain aspects of service provision and reconstruction (Safapour et al., 2021). Allocation of financial resources to disaster recovery efforts has always been limited but may become even more so in future; the significant increase in climate-induced disasters, for example, has led to decreased donations and funding for organizations to provide psychological support services. While the importance of mental health support is becoming more mainstreamed, the focus on this service in a post-disaster context continues to be an afterthought. Because of this, there is often limited funding available to provide psychological support services.

Future Directions and Conclusion

Disaster-focused prevention and treatment interventions are critical to maintaining and restoring the well-being of children, youth, and families. Despite challenges of research and service delivery in emergency contexts, significant strides have been made in provision of psychological support to survivors. Intervention modalities, for example, have evolved from focusing primarily on treatment for individuals with clinical manifestations of distress to universal approaches for all survivors. Continued research on the efficacy of interventions, however, is essential to understand the long-term benefits of these services during both the immediate disaster and longer-term recovery. Additionally, as programs scale up and become more accessible in emergency contexts, it is vital to examine how they are being implemented. Many interventions are evidence-based. However, adaptations are often made by service providers based on the needs of their clients. Ongoing assessment of ad hoc adaptations and barriers, as well as facilitators, to program delivery would continue to inform how to provide effective services in post-disaster contexts.

There also is a shortage of interventions that address the psychological needs of families in post-disaster contexts. Many programs are designed to treat the individual child or adult, but do not target family factors that are essential to post-disaster adaptation. Family-focused interventions during and after a disaster may include psychoeducation on risk and protective factors for post-disaster adjustment, activities to increase parent/child connectedness, and tools to reduce physical and emotional stress responses.

As disasters continue to increase, affecting the global population, accessible evidence-based psychological services are essential to support post-disaster recovery. Many lessons have been learned over the past 20 years. However, continued intervention research and development are critical to help individuals and families adapt to and recover from these emergencies.

References

Aarons, G. A., Green, A. E., Palinkas, L. A., Self-Brown, S., Whitaker, D. J., Lutzker, J. R., Silovsky, J. F., Hecht, D. B., & Chaffin, M. J. (2012). Dynamic adaptation process to implement an evidence-based child maltreatment intervention. *Implementation Science, 7*(1), 1–9. https://doi.org/10.1186/1748-5908-7-32

Ager, A., Stark, L., Akesson, B., & Boothby, N. (2010). Defining best practice in care and protection of children in crisis-affected settings: A Delphi study. *Child Development, 81*(4), 1271–1286. https://doi.org/10.1111/j.1467-8624.2010.01467.x

Ahmad, A., & Sundelin-Wahlsten, V. (2008). Applying EMDR on children with PTSD. *European Child & Adolescent Psychiatry, 17*(3), 127–132. https://doi.org/10.1007/s00787-007-0646-8

Alexander, L., Carpenter, L., Simpson, J., & Gibbs, L. (2021). *Journey of hope evaluation*. Save the Children.

Amaratunga, C. A., & O'Sullivan, T. L. (2006). In the path of disasters: Psychosocial issues for preparedness, response, and recovery. *Prehospital and Disaster Medicine, 21*(3), 149–153. https://doi.org/10.1017/s1049023x00003605

Ardagh, M. W., Richardson, S. K., Robinson, V., Than, M., Gee, P., Henderson, S., Khodaverdi, L., McKie, J., Robertson, G., & Schroeder, P. P. (2012). The initial health-system response to the earthquake in Christchurch, New Zealand, in February, 2011. *The Lancet, 379*(9831), 2109–2115. https://doi.org/10.1016/S0140-6736(12)60313-4

Arshad, M., Mughal, M. K., Giallo, R., & Kingston, D. (2020). Predictors of child resilience in a community-based cohort facing flood as natural disaster. *BMC Psychiatry, 20*(1), 1–10. https://doi.org/10.1186/s12888-020-02944-y

Bandura, A. (1998). Health promotion from the perspective of social cognitive theory. *Psychology and Health, 13*(4), 623–649. https://doi.org/10.1080/08870449808407422

Becker-Blease, K. A., Turner, H. A., & Finkelhor, D. (2010). Disasters, victimization, and children's mental health. *Child Development, 81*(4), 1040–1052. https://doi.org/10.1111/j.1467-8624.2010.01453.x

Benight, C. C., & Bandura, A. (2004). Social cognitive theory of posttraumatic recovery: The role of perceived self-efficacy. *Behaviour Research and Therapy, 42*(10), 1129–1148. https://doi.org/10.1016/j.brat.2003.08.008

Blanchet-Cohen, N., & Nelems, R. (2013). A child-centered evaluation of a psychosocial program: Promoting children's healing, safety and well-being in post-disaster contexts. *Children, Youth and Environments, 23*(1), 23–42. https://doi.org/10.7721/chilyoutenvi.23.1.0023

Boscarino, J. A., Hoffman, S. N., Adams, R. E., Figley, C. R., & Solhkhah, R. (2014). Mental health outcomes among vulnerable residents after hurricane Sandy: Implications for disaster research and planning. *American Journal of Disaster Medicine, 9*(2), 107–120. https://doi.org/10.5055/ajdm.2014.0147

Brown, R. C., Witt, A., Fegert, J. M., Keller, F., Rassenhofer, M., & Plener, P. L. (2017). Psychosocial interventions for children and adolescents after man-made and natural disasters: A meta-analysis and systematic review. *Psychological Medicine, 47*(11), 1893–1905. https://doi.org/10.1017/S0033291717000496

Cadamuro, A., Versari, A., Vezzali, L., & Trifiletti, E. (2016). Preventing the detrimental effect of posttraumatic stress in young children: The role of theory of mind in the aftermath of a natural disaster. *European Journal of Developmental Psychology, 13*(1), 52–66. https://psycnet.apa.org/doi/10.1080/17405629.2015.1055240

Cadamuro, A., Birtel, M. D., Di Bernardo, G. A., Crapolicchio, E., Vezzali, L., & Drury, J. (2021). Resilience in children in the aftermath of disasters: A systematic review and a new perspective on individual, interpersonal, group, and intergroup level factors. *Journal of Community & Applied Social Psychology, 31*(3), 259–275. https://doi.org/10.1002/casp.2500

Carroll, C., Patterson, M., Wood, S., Booth, A., Rick, J., & Balain, S. (2007). A conceptual framework for implementation fidelity. *Implementation Science, 2*(1), 1–9. https://doi.org/10.1186/1748-5908-2-40

Chen, S., Bagrodia, R., Pfeffer, C. C., Meli, L., & Bonanno, G. A. (2020). Anxiety and resilience in the face of natural disasters associated with climate change: A review and methodological critique. *Journal of Anxiety Disorders, 76*. https://doi.org/10.1016/j.janxdis.2020.102297

Cicchetti, D. (2010). Resilience under conditions of extreme stress: A multilevel perspective. *World Psychiatry, 9*(3), 145–154. https://doi.org/10.1002/j.2051-5545.2010.tb00297.x

Cobham, V. E., McDermott, B., Haslam, D., & Sanders, M. R. (2016). The role of parents, parenting and the family environment in children's post-disaster mental health. *Current Psychiatry Reports, 18*(6), 53. https://doi.org/10.1007/s11920-016-0691-4

Cohen, J. A., & Mannarino, A. P. (2008). Trauma-focused cognitive behavioural therapy for children and parents. *Child and Adolescent Mental Health, 13*(4), 158–162. https://doi.org/10.1111/j.1475-3588.2008.00502.x

Collins, S. E., Clifasefi, S. L., Stanton, J., The Leap Advisory Board, Straits, K. J., Gil-Kashiwabara, E., Rodriguez Espinosa, P., Nicasio, A. V., Andrasik, M. P., Hawes, S. M., Miller, K. A., Nelson, L. A., Orfaly, V. E., Duran, B. M., & Wallerstein, N. (2018). Community-based participatory research (CBPR): Towards equitable involvement of community in psychology research. *American Psychologist, 73*(7), 884–898. https://doi.org/10.1037/amp0000167

DiGangi, J. A., Gomez, D., Mendoza, L., Jason, L. A., Keys, C. B., & Koenen, K. C. (2013). Pretrauma risk factors for posttraumatic stress disorder: A systematic review of the literature. *Clinical Psychology Review, 33*(6), 728–744. https://doi.org/10.1016/j.cpr.2013.05.002

Dyregrov, A., Yule, W., & Olff, M. (2018). Children and natural disasters. *European Journal of Psychotraumatology, 9*(Suppl 2), 1–3. https://doi.org/10.1080/20008198.2018.1500823

Fazel, M., Stratford, H. J., Rowsell, E., Chan, C., Griffiths, H., & Robjant, K. (2020). Five applications of narrative exposure therapy for children and adolescents presenting with post-traumatic stress disorders. *Frontiers in Psychiatry, 11*, 19. https://doi.org/10.3389/fpsyt.2020.00019

Felix, E., Kaniasty, K., You, S., & Canino, G. (2016). Parent–child relationship quality and gender as moderators of the influence of hurricane exposure on physical health among children and youth. *Journal of Pediatric Psychology, 41*(1), 73–85. https://doi.org/10.1093/jpepsy/jsv038

Felix, E., Rubens, S., & Hambrick, E. (2020). The relationship between physical and mental health outcomes in children exposed to disasters. *Current Psychiatry Reports, 22*(7). https://doi.org/10.1007/s11920-020-01157-0

Gilbert, R., Abel, M. R., Vernberg, E. M., & Jacobs, A. K. (2021). The use of psychological first aid in children exposed to mass trauma. *Current Psychiatry Reports, 23*(9), 1–9. https://doi.org/10.1007/s11920-021-01270-8

Grolnick, W. S., Schonfeld, D. J., Schreiber, M., Cohen, J., Cole, V., Jaycox, L., Lochman, J., Pfefferbaum, B., Ruggiero, K., Wells, K., Wong, M., & Zatzick, D. (2018). Improving adjustment and resilience in children following a disaster: Addressing research challenges. *American Psychologist, 73*(3), 215–229. https://doi.org/10.1037/amp0000181

Gutman, G. M., & Yon, Y. (2014). Elder abuse and neglect in disasters: Types, prevalence and research gaps. *International Journal of Disaster Risk Reduction, 10*, 38–47. https://doi.org/10.1016/J.IJDRR.2014.06.002

Hackbarth, M., Pavkov, T. W., Wetchler, J., & Flannery, M. (2012). Natural disasters: An assessment of family resiliency following Hurricane Katrina. *Journal of Marital and Family Therapy, 38*(2), 340–351. https://doi.org/10.1111/j.1752-0606.2011.00227.x

Higgen, S., Mueller, J. T., & Mösko, M. (2021). Universal mental health interventions for young students in adverse environments – A systematic review of evaluated interventions. *Child and Adolescent Mental Health*. https://doi.org/10.1111/camh.12493

Jia, Z., Tian, W., Liu, W., Cao, Y., Yan, J., & Shun, Z. (2010). Are the elderly more vulnerable to psychological impact of natural disaster? A population-based survey of adult survivors of the 2008 Sichuan earthquake. *BMC Public Health, 10*(1), 172. https://doi.org/10.1186/1471-2458-10-172

Jovanovic, A. A., Aleksandric, B. V., Dunjic, D., & Todorovic, V. S. (2004). Family hardiness and social support as predictors of post-traumatic stress disorder. *Psychiatry, Psychology and Law, 11*(2), 263–268. https://doi.org/10.1375/pplt.2004.11.2.263

Kouadio, I. K., Aljunid, S., Kamigaki, T., Hammad, K., & Oshitani, H. (2012). Infectious diseases following natural disasters: Prevention and control measures. *Expert Review of Anti-Infective Therapy, 10*(1), 95–104. https://doi.org/10.1586/eri.11.155

Kousky, C. (2016). Impacts of natural disasters on children. *The Future of Children, 26*, 73–92. https://files.eric.ed.gov/fulltext/EJ1101425.pdf

Kronenberg, M. E., Hansel, T. C., Brennan, A. M., Osofsky, H. J., Osofsky, J. D., & Lawrason, B. (2010). Children of Katrina: Lessons learned about postdisaster symptoms and recovery patterns. *Child Development, 81*(4), 1241–1259. https://doi.org/10.1111/j.1467-8624.2010.01465.x

La Greca, A. M., Lai, B. S., Joormann, J., Auslander, B. B., & Short, M. A. (2013). Children's risk and resilience following a natural disaster: Genetic vulnerability, posttraumatic stress, and depression. *Journal of Affective Disorders, 151*(3), 860–867. https://doi.org/10.1016/j.jad.2013.07.024

Lazarus, R. S., & Folkman, S. (1984). *Stress, appraisal, and coping*. Springer.

Lewey, J. H., Smith, C. L., Burcham, B., Saunders, N. L., Elfallal, D., & O'Toole, S. K. (2018). Comparing the effectiveness of EMDR and TF-CBT for children and adolescents: A meta-analysis. *Journal of Child & Adolescent Trauma, 11*(4), 457–472. https://doi.org/10.1007/s40653-018-0212-1

Masten, A. S. (2013). Risk and resilience in development. In P. D. Zelazo (Ed.), *The Oxford handbook of developmental psychology* (*Self and other*) (Vol. 2, pp. 579–607). Oxford University Press. https://doi.org/10.1093/oxfordhb/9780199958474.013.0023

Masten, A. S. (2021). Resilience of children in disasters: A multisystem perspective. *International Journal of Psychology, 56*(1), 1–11. https://doi.org/10.1002/ijop.12737

Masten, A. S., & Barnes, A. J. (2018). Resilience in children: Developmental perspectives. *Children, 5*(7), 98. https://doi.org/10.3390/children5070098

McDonald-Harker, C., Drolet, J. L., Sehgal, A., Brown, M. R. G., Silverstone, P. H., Brett-MacLean, P., & Agyapong, V. I. O. (2021). Social-ecological factors associated with higher levels of resilience in children and youth after disaster: The importance of caregiver and peer support. *Frontiers in Public Health, 9*, 682634. https://doi.org/10.3389/fpubh.2021.682634

Mooney, M., Tarrant, R., Paton, D., Johnston, D., & Johal, S. (2020). The school community contributes to how children cope effectively with a disaster. *Pastoral Care in Education, 39*, 24–47. https://doi.org/10.1080/02643944.2020.1774632

Parkinson, D., & Zara, C. (2013). The hidden disaster: Domestic violence in the aftermath of natural disaster. *Australian Journal of Emergency Management, 28*(2), 28–35.

Peek, L. (2008). Children and disasters: Understanding vulnerability, developing capacities, and promoting resilience—An introduction. *Children, Youth and Environments, 18*(1), 1–29.

Pfefferbaum, B., Jacobs, A. K., Griffin, N., & Houston, J. B. (2015). Children's disaster reactions: The influence of exposure and personal characteristics. *Current Psychiatry Reports, 17*(7), 56. https://doi.org/10.1007/s11920-015-0598-5

Powell, T., & Bui, T. (2016). Supporting social and emotional skills after a disaster: Findings from a mixed methods study. *School Mental Health, 8*(1), 106–119. https://psycnet.apa.org/doi/10.1007/s12310-016-9180-5

Powell, T., & Leytham, S. (2014). Building resilience after a natural disaster: An evaluation of a parental psycho-educational curriculum. *Australian Social Work, 67*(2), 285–296. https://doi.org/10.1080/0312407X.2014.902981

Powell, T., Wegmann, K. M., & Backode, E. (2021). Coping and post-traumatic stress in children and adolescents after an acute onset disaster: A systematic review. *International Journal of Environmental Research and Public Health, 18*(9), 4865. https://doi.org/10.3390/ijerph18094865

Robjant, K., & Fazel, M. (2010). The emerging evidence for narrative exposure therapy: A review. *Clinical Psychology Review, 30*(8), 1030–1039. https://doi.org/10.1016/j.cpr.2010.07.004

Rufat, S., Tate, E., Burton, C. G., & Maroof, A. S. (2015). Social vulnerability to floods: Review of case studies and implications for measurement. *International Journal of Disaster Risk Reduction, 14*, 470–486. https://doi.org/10.1016/j.ijdrr.2015.09.013

Safapour, E., Kermanshachi, S., & Pamidimukkala, A. (2021). Post-disaster recovery in urban and rural communities: Challenges and strategies. *International Journal of Disaster Risk Reduction, 64*. https://doi.org/10.1016/j.ijdrr.2021.102535

Sanchez, A. L., Cornacchio, D., Poznanski, B., Golik, A. M., Chou, T., & Comer, J. S. (2018). The effectiveness of school-based mental health services for elementary-aged children: A meta-analysis. *Journal of the American Academy of Child & Adolescent Psychiatry, 57*(3), 153–165. https://doi.org/10.1016/j.jaac.2017.11.022

Save the Children. (2014). *Journey of hope: Cultural adaptation guidance* [Unpublished manuscript].

Save the Children. (2022, January). *Protecting America's kids: Social and emotional learning programs*. https://www.savethechildren.org/us/what-we-do/protection/child-protection

Shultz, J. M., & Forbes, D. (2014). Psychological first aid: Rapid proliferation and the search for evidence. *Disaster Health, 2*(1), 3–12. https://doi.org/10.4161/dish.26006

Vu, L., & VanLandingham, M. J. (2012). Physical and mental health consequences of Katrina on Vietnamese immigrants in New Orleans: A pre-and post-disaster assessment. *Journal of Immigrant and Minority Health, 14*(3), 386–394. https://doi.org/10.1007/s10903-011-9504-3

Wang, C.-W., Chan, C. L., & Ho, R. T. H. (2013). Prevalence and trajectory of psychopathology among child and adolescent survivors of disasters: A systematic review of epidemiological studies across 1987–2011. *Social Psychiatry and Psychiatric Epidemiology, 48*(11), 1697–1720. https://doi.org/10.1007/s00127-013-0731-x

Weare, K., & Nind, M. (2011). Mental health promotion and problem prevention in schools: What does the evidence say? *Health Promotion International, 26*(Suppl 1), i29–i69. https://doi.org/10.1093/heapro/dar075

Wolmer, L., Hamiel, D., & Laor, N. (2011). Preventing children's posttraumatic stress after disaster with teacher-based intervention: A controlled study. *Journal of the American Academy of Child & Adolescent Psychiatry, 50*(4), 340–348. https://doi.org/10.1016/j.jaac.2011.01.002

Zhou, D. (2018). The role of family in children with PTSD after natural disasters. *Journal of Biosciences and Medicines, 6*(12), 111–127. https://doi.org/10.4236/jbm.2018.612011

Part II
Climate Change

Chapter 3
Climate Change-Related Demographic and Health Research: Data and Approaches

Audrey Dorélien and Kathryn Grace

Climate change is associated with changes in long-term average temperature and precipitation conditions, as well as extreme events such as droughts, floods, and heat waves. When classifying and examining these events, researchers pay particular attention to their frequency, duration, and severity (Grace et al., 2021; Meehl & Tebaldi, 2004). Shifts in seasonal conditions, such as shorter or delayed rainy seasons or greater variability in rainfall conditions during the rainy season, are also associated with climate change. In the context of global warming, individuals may, in their lifetimes, experience all or some combination of these three conditions associated with climate change: a gradual (i.e., minimally noticeable) increase in average warming and drying conditions, repeated extreme weather events, or small but significant shifts in seasonal precipitation and temperature.

Exposure to these different conditions may have significant adverse impacts on health and demographic outcomes that are evident immediately after exposure. While in other cases, the effects may not be observed until years later (McMichael, 2013; McMichael et al., 2006). Some adverse outcomes can impact an individual's entire life (e.g., stunting), while others result in acute and resolvable outcomes (e.g., wasting; Baker & Anttila-Hughes, 2020; Isen et al., 2017; Thiede & Strube, 2020). Understanding the spatial footprint of an exposure in order to identify the communities impacted by the conditions is an additional dimension to consider. Because rainfall is highly spatially variable, droughts, for example, can occur in a very small

A. Dorélien (✉)
Hubert H. Humphrey School of Public Affairs, University of Minnesota, Minneapolis, MN, USA
e-mail: dorelien@umn.edu

K. Grace
Department of Geography, Environment, and Society, University of Minnesota, Minneapolis, MN, USA
e-mail: klgrace@umn.edu

S. E. Ortiz et al. (eds.), *Environmental Impacts on Families*, National Symposium on Family Issues 12, https://doi.org/10.1007/978-3-031-22649-6_3

area that includes a single village or in several geographically linked[1] towns or villages. In contrast, elevated temperatures can impact an entire country or large subregion of a country (Hill et al., 2019).

Beyond spatial and temporal considerations, risk or vulnerability to adverse conditions increases or decreases in response to a range of interrelated demographic, social, economic, political, and environmental factors (Brown et al., 2021; Watts et al., 2015). Varying individual- and household-level factors such as age, marital status, electricity status, and wealth (among others) mean that risks and vulnerabilities to adverse impacts are not equally distributed (Grace et al., 2020; Hunter et al., 2021). In practice, individual-level heterogeneity makes linking climate change to family experiences and outcomes complex (Segnon et al., 2021). Beyond the challenge of classifying a "family" or "household" structure for research purposes (Randall et al., 2015), families within the same community facing the same exposures may experience very different economic and health outcomes shaped by how they live and what they have access to within that community (Parsons, 2019; Rao et al., 2019; Segnon et al., 2021). Within families, individual members may experience differential vulnerability to the same climatic event because of their life course stage, intra-household bargaining power, and gender (Grace et al., 2020; Rao et al., 2020).

Given the complexity of the exposure factors and the variability of ways that families, households, and individuals cope and respond, understanding the processes that link climate change conditions to health and demographic outcomes requires detailed, diverse, and innovative data, theory, and analytic tools. Combining these data, theory, and tools to advance climate change research represents a new type of integrated scientific approach that engages both physical and social sciences.

In this chapter, we describe the key features to consider when linking climate change data with spatial socio-demographic and health data to better understand the processes of climate vulnerabilities and associated health and policy interventions. We also describe characteristics from several popular and publicly available health datasets that are important for users to consider. We specifically target this chapter towards low-income country settings because this is our primary area of research, but we note that much of what we describe and explore is relevant to wealthier settings as well. We organize this chapter by first establishing the importance of quantification and measurement in this domain. Next, we describe the importance of properly linking, in space and time, spatial demographic and health data to climate and environmental data to support quantitative analysis. We then discuss the importance of considering social factors and mitigation strategies (including migration) that may reduce or exacerbate climate change impacts. Afterwards, we highlight two projects that apply the advice to study the impact of exposure to hot

[1] "Linked" can refer to the spatial proximity of a set of villages or it can refer to some other factor that connects villages (e.g., roads and trade partners).

temperatures, rainfall, and malarious conditions on maternal and child health[2] in sub-Saharan Africa. We conclude with a discussion of the gaps and opportunities for family researchers interested in studying climate change impacts on health and well-being.

Defining Climate Change Measures for Demographic-Health Research

Clearly describing strategies for quantifying and measuring climate change as it relates to demographic, health, and social science impacts is a vital step in interdisciplinary climate change research (Grace et al., 2020; Hunter et al., 2021). In fact, discussions of measurement and event classification continue to emerge as scholars work to advance population-environment research and seek to develop a cohesive approach or summary of evidence (Green et al., 2019; Hill et al., 2019; Phalkey et al., 2015; Xu et al., 2016). For example, the term "drought" is defined in different ways depending on the setting, the community of impact, and the primary discipline of the research team conducting the study (see also Hill et al., 2019; Sohnesen, 2020). By definition, a drought is a period of time when there is less rainfall than what is normally experienced in a particular location (see USGS). The period could be days, weeks, years, or some other unit of time. In existing research, we may see a drought measured using seasonal rainfall totals (or z-scores/anomalies), and we may see this measured at different spatial scales. We will discuss in more detail later.

Local Context Matters in Measurement of Climate Change

The same climatic condition can have distinct impacts based on setting. Take the case of a delayed rainy season in Mali versus Tajikistan. A farmer in a more arid, rainfed setting where the rainy season is relatively short and defined (e.g., Mali) may face seriously reduced seasonal agricultural yields if the rainy season is shortened or delayed (e.g., Shukla et al., 2021). Reduced yields may occur in this situation even if the total seasonal precipitation value is not significantly different from the long-term norm. Therefore, the start of the season or length of season measures may more accurately capture rainfall conditions associated with reduced agricultural yields, versus seasonal totals (or z-scores/anomalies). These sub-seasonal measures capture below average rainfall (or drought conditions) during a relatively short period of time (days or weeks for communities in Mali). However, when

[2] Variables commonly used to measure different dimensions of maternal and child health include infant mortality, maternal mortality, birth weight, child malnutrition (as measured by height-for-age, weight-for-age, and weight-for-height), fertility/childbearing (e.g., spacing, timing, fecundity, goals, and total number of live births), and family planning use/intentions.

considering the season as a whole, there may be little indication of a drought. Moreover, researchers and farmers might not necessarily refer to a period of delayed seasonal rainfall as a drought period. Rather, they may classify this as a late-starting or shorter-than-average season. In contrast, in settings where different forms of irrigation are used because the primary rainy season does not necessarily overlap with the growing season (e.g., Tajikistan), small seasonal shifts in the timing and level of rainfall may have less of an impact on yields. Also, the source of water may be somewhat distant from the agricultural plot, and the process of storing and transporting water to the agricultural field may result in additional measurement challenges.

Further complicating measurement of drought conditions is the fact that even within a community, climate change can have heterogeneous effects. A range of adaptation strategies may be used in a community to modify the relationship between rainfall and its impact. As a result, families in the same community (exposed to the same environmental conditions) may face very dissimilar agricultural outcomes. For example, crops grown in the same community may require distinct growing conditions for optimal yields; some farmers may rely on drought-resistant seeds or on fertilizers, and individuals may have differential access to irrigation. This kind of variability means that even when farmers experience the same external environmental conditions, the impacts may be unalike because of these types of farmer-specific factors. Furthermore, families' differential access to social safety nets and insurance influences how the farmers cope with droughts – thereby influencing demographic and health measures that analysts may be interested in. All of this variability makes it extremely challenging to apply a single quantitative definition of drought that is applicable across settings and relevant to a range of individuals.

We describe the various complications with drought measurement as they relate to agricultural producers in low-income settings to highlight the many decisions an analyst must make when evaluating how climate change conditions impact humans. As demonstrated by the numerous considerations and context-specific factors that contribute to possible definitions of drought, informed and clearly identified spatial and temporal measurement of such climate exposures is critical. Further, these exposures should be thoughtfully, appropriately, and explicitly linked to the particular outcome of interest. Multiple definitions consequently generate various measurement and categorization strategies and ultimately may support various quantitative results (Phalkey et al., 2015; Sohnesen, 2020). Thus, it is vital that those involved in modeling and quantifying climate change risks clearly describe their research design, including the spatial or temporal details of the data in the theoretical and quantitative model.

Applied scientists in diverse disciplines (e.g., nutrition, agriculture, economics, ecology) often examine the relationship between climate change and downstream outcomes experienced by humans. In these studies, a range of climate measures that consider multiple time periods or spatial aggregations are examined

Table 3.1 Common environmental measures used in climate change research

Climate change measures	Spatial detail	Temporal aggregation
Temperature: surface, humidity, wet-bulb	Typically, in raster form ranging from around 5 km to 50 km resolution; can also come from meteorological stations	Z-scores using a 50-year mean; counts/bins of hot days; average recent conditions
Precipitation: rainfall, snowfall	Finer scale is often better but an understanding of snow melt, irrigation, and water use is key to capture conditions of relevant places. Existing data is often in raster form with a 5 km resolution	Varies greatly; may want to consider variability, dry days, start of season, total seasonal conditions
Sea level rise: coastal flooding and storm surges, surface runoff data, flood extent	Runoff data fine resolution 0.25*0.25 grids; flood extent shapefiles	Hourly or monthly for surface runoff data; static flood shapefiles
Shifts in seasonal conditions relevant to disease spread (e.g., conditions suited to malaria or meningitis spread) or to agricultural production (e.g., start of season, yield, and others)	Vegetation measures are often fine scale (< 5 km) and need to be aggregated or linked to individuals in relevant ways; disease conditions can be created using composites of other factors	Helpful to consider key growing or hunger seasons and summarize agricultural production or food availability during those periods. With disease, may want to consider how conditions vary over time and interact with population-level immunity

and compared as a key component of advancing climate change science (Table 3.1). Demography and population research have less fully engaged with the wide range of measures capturing various dimensions of climate change impacts. Scholars frequently rely on anomalies (or deviations from some mean) with little explicit discussion about how an anomaly would trickle down to impact an individual's life. The implications of the choice of a particular time frame are also rarely explored with some scholars using a long-term mean and others using shorter-term means. The decision-making around a particular time period as a baseline to compare yearly conditions must be described and justified with relation to the setting and the research question. There are research questions for which it may make sense to consider variations in a given year as compared to the 60-year mean, while in other settings and for other questions, it may be more appropriate to consider how conditions have changed over a shorter (e.g., 5 year) period of time. For example, considering long-term conditions may provide insight into the food and agricultural system and associated infrastructure of an area (e.g., broadly identifying pastoral versus cropping zones), while considering more short-term conditions may be useful for examining short-term adaptation and coping strategies of farming households.[3] These kinds of considerations help researchers and policymakers to refine their understanding of the processes that link climate change to health outcomes, and by clearly describing measurement approaches, scientists may uncover opportunities to re-examine assumptions, refine measures, and advance research.

Spatial and Temporal Scales: Linking Socio-Demographic Datasets to Climate Data

The most common source of data on population, health, and well-being is household surveys. Data can also come from censuses or from vital records, though censuses are not as frequent (collected every 10 years at best) and often are not as detailed. Since household survey datasets rarely include data on climate change, the datasets must have spatial information (identifiers), such as the latitude and longitudinal coordinates of a village center, and temporal identifiers such as dates, months, and years. Spatial and temporal identifiers allow researchers to link the surveyed individuals, households, or communities to specific climate change exposures at a specific time.

[3] In a context of climate change, recent conditions (e.g., those within the previous 5–10 years) may be notably different from the "long-term mean," and families and individuals likely have adapted to a new norm requiring attention to more recent conditions rather than long-term trends.

Spatial Scales

The finest spatial data that are available in household surveys are geocoded points, also known as latitude and longitude coordinates. However, the analysis of geocoded points often requires the use of a spatial buffer to account for the fact that the true spatial location has been randomly spatially displaced up to a certain distance to protect survey respondent privacy. This is the case for the Demographic and Health Survey (DHS) and similar household surveys described in the next section. Figure 3.1a shows a geocoded point (sampling cluster) surrounded by 10 km buffer. Publicly available protocols describing how these data were collected and altered for public release provides valuable information on sources of measurement error and allows the analyst to appropriately merge it with climate data.

Sub-national administrative units such as city or district names can also be used to link the demographic and health data with climate data, because they are associated with known spatial files. However, using coarser administrative units requires the spatial aggregation of climate data (Fig. 3.1b). A finer spatial scale means more precision and is often preferable; for example, geocoded points would be preferable to city names (administrative unit level 3) and city names preferable to county names (administrative unit level 2). More precise spatial reference points (i.e., locations) allow researchers to merge the demographic data with high spatial resolution climate data and thereby more accurately capture the specific climate exposure.

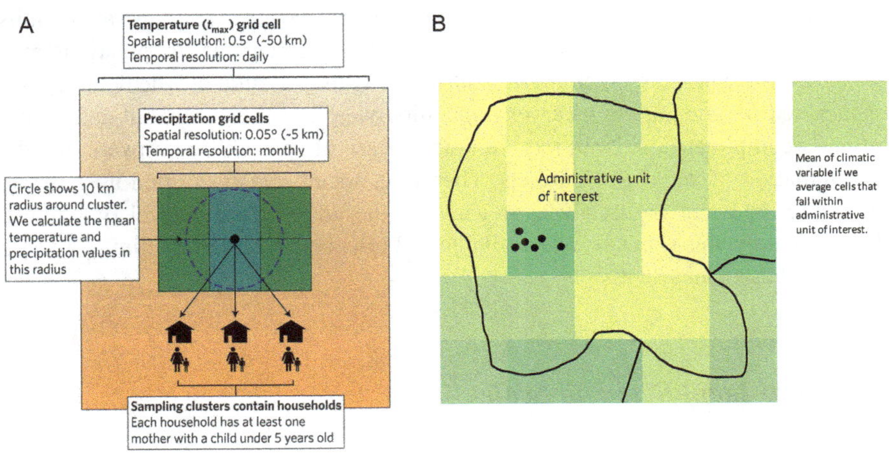

Fig. 3.1 Challenges of merging data with different spatial scales

Note*:* Panel A shows challenges of merging geocoded points with climate data. Panel B illustrates the challenges of linking climate data to individuals/households when the only spatial identifier is the administrative unit. The values of the climatic variables assigned to the participants within that administrative unit may not be representative of their actual climate exposure, especially if participants are not evenly distributed within the administrative unit. Furthermore, analysts must determine how to appropriately aggregate the climatic variables to the administrative unit.

(*Source*: Panel A is adapted from Grace, 2017)

This approach enables researchers to exploit the fine spatial detail often found in climate data and is particularly important for climate change measures that are heterogeneous across small spatial scales such as precipitation. If spatial scales are well matched, two neighboring geocoded points are less likely to be assigned the same climatic exposure. Alternatively, finer scale climate data is not always necessary, and the computational demands of very high-resolution data can quickly reduce the benefits. Satellite-based remotely sensed data of land cover, for example, exists at very fine spatial scales (e.g., <3 m). However, this level of spatial detail is rarely required in a climate-health analysis and would require an analyst to aggregate the climate data in some way to align with the human health process of interest. Thus, in general, researchers must carefully and deliberately consider the spatial and temporal scales of the human data and the climate data and merge them in ways that reflect the realities of the data but that also can best address the research goal.

Researchers often face challenges when existing and available data have mismatched spatial scales. For example, linking a high spatial resolution demographic and health dataset with spatially coarse climate data means that the values of the climatic exposures assigned to individuals may not be representative of that individual's true exposure to climate. Furthermore, this reduces the amount of variation in the climate variables of interest available for the analyst to exploit in their analyses. Barriers also arise when the climatic data is available at a finer spatial scale compared to the demographic data. For instance, if a dataset contains only the spatial information of a participants' county of residence, it may be difficult to assign the participants to the correct exposure even if fine resolution climate data is available (Fig. 3.1b). As previously mentioned, using coarser administrative units requires spatial aggregation of climate data, but settlements and populations may not be evenly distributed within the county, raising the question of whether to aggregate based on geographic mean or population-weighted mean. Spatial data wrangling is complicated and can require both the use of specialized software and the manipulation of very large datasets. Therefore, social scientists often use climate exposure data that is already aggregated to the administrative level of interest. Unfortunately, this is not always available and is a notable barrier to climate-related research for social scientists.

Temporal Identifiers and Scales

In addition to linking socio-demographic data with climate data in space, it is also necessary to link in time. Consequently, both types of data must have temporal identifiers. In a socio-demographic dataset, it is useful to have the following time variables/markers:

1. Length of time in the community: This allows the analyst to identify the population that was exposed in a given community from among those who currently reside there. In other words, an individual who participates in the survey may not

have been in residence in the community during a particular period of time when a climate-related event occurred. The length of time information was not available for all the datasets used by Grace et al. (2021) in their study of environmental conditions and birth weight in Mali. However, to ensure the robustness of their results, they estimated models for the two different populations – the sample with matched exposures based on length of time at current residence and the sample with no information about length of time in current residence. Approaches such as this can be used in the case that there is not sufficient information on residence. Alternatively, researchers could employ external estimates of mobility into and out of an area for a general sense of how much the population might be changing. In some cases, as in the aftermath of hurricanes, forest fires, or other natural disasters, there may be dramatic changes in the composition of the population, and these factors must be considered in the research design (Fussell et al., 2010; Groen & Polivka, 2010).

2. Birth date (ideally day, month, and year): This allows linking climatic variables to each day, month, or year of an individual's life course (e.g., conception, birth, key growth stages, and others). Individuals may reside in the same location and be exposed to the same climatic conditions. However, that exposure would be linked to the individuals at different ages in their life course. For example, month and year of a birth can be used to link the in utero period and each month of infant life to the number of days per month within specific temperature bins[4] to study the impact of hot/humid days on infant mortality (Geruso & Spears, 2018). "Wet-bulb" temperature was specifically considered as a way to potentially capture the "feels like" conditions. Figure 3.2 illustrates how birth date can be used to link climatic data. Some climatic exposures are of a very short duration. Therefore, if the births are recorded in month-year format instead of as exact dates, the analyst may incorrectly assign an individual to a specific exposure. For instance, exact birth dates are needed to determine whether individuals were exposed to a three-day heat wave during their month of birth. For those interested in the impacts of in utero exposure, ideal data would include conception date/gestational duration. Without specific information on the date of birth, the analyst will have to assign a birth date to a child at random (within a given month and year). In this case, the analyst should consider the robustness of their results to different data assignments and describe their choices and assumptions.
3. Interview date: Some measures of health and well-being vary over time. These include health factors such as diarrhea occurrence or fever, as well as household income and employment status. Other measures are influenced by an individual's exact age, such as height-for-age (stunting) or weight-for-height (wasting). For this reason, interview dates matter. For instance, in the Demographic and Health Surveys, interview date is linked with information on temperature and

[4] Binning temperature or counting the days within a range of temperatures (e.g., 71–80F, 81–90F, 91–100F) is an approach commonly used by economists when studying temperature and health (see also Deschênes et al., 2009; Isen et al., 2017).

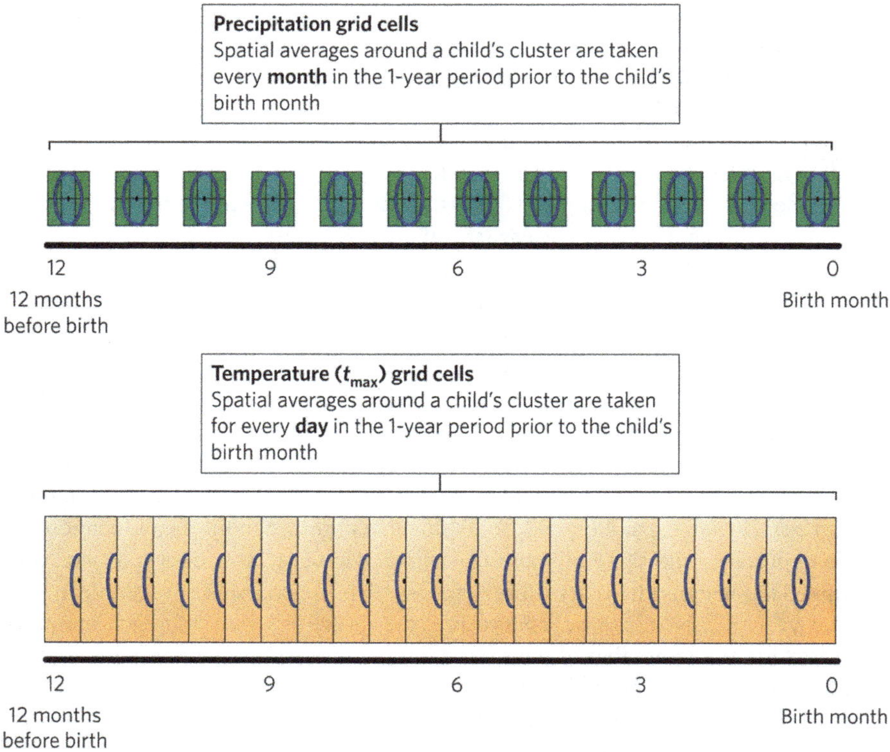

Fig. 3.2 Temporal identifiers can be linked to climate data with varying temporal scales

Note*:* Illustrates how birth date can be used to temporally link socio-demographic data with climatic variables. Climate data is also available at different temporal scales. In this example the precipitation data is at the monthly level, while the temperature data is at the daily level.

(*Source:* Adapted from (Grace, 2017))

humidity to study the impact of weather on interviewer productivity (LoPalo, 2020). Surveys can be conducted over multiple months. Therefore, information on interview date is needed to control for seasonally varying factors; and interview year is needed to control for time trends when pooling multiple surveys together.

4. Dates of key life course events: Birth date and dates of other key life events may be required for research. Many events may be associated in different ways with climate change and health including school start/stop, union formation/dissolution, employment type/status, migration events, hospital stays, and date of death. In some DHS surveys, specific months of union formation/dissolution are included and can potentially be linked with climate conditions to determine, for example, if a drought event (resulting in a reduction in household food and income resources) impacts the likelihood of union formation.

Many of these temporal identifiers were used to study the effect of increased temperatures on child (ages 1–5 years) malnutrition in sub-Saharan Africa. Specifically, the interview date was used to study the impact of increased temperatures in the month of the survey and the year leading up to the survey, while the children's birth dates were used to study the effects of increased temperatures during the children's lifetimes (Baker & Anttila-Hughes, 2020). A similar health dataset and approach to space-time matching were used to evaluate the impact of drought on child health demonstrating that when precipitation deviates from long-term norms (positively or negatively), children's health suffers (Cooper et al., 2019).

Climate data is typically available at very fine temporal detail (e.g., daily or even sub-daily). Therefore, there are fewer data constraints with the climate data. However, it is vital to reflect on how to operationalize the exposure in a way that corresponds to the research question. Temporal aggregation depends on the mechanism of interest (Grace et al., 2020). Rarely is it appropriate to link the temporal identifier in a socio-demographic dataset to one specific value in the climatic data such as precipitation on survey date. As a rare example that aligns with the broader area of time-use scholarship, see LoPalo (2020). Often, the climatic exposure is a composite measure. Daily precipitation may not be as useful as average monthly precipitation or 3-month cumulative rainfall total, particularly if interested in assessing agricultural conditions.

Linking Demographic-Health Surveys with Climate Datasets

The effects of climate change are unequally distributed with some families being more susceptible and vulnerable. Exposure to climate change risks is exacerbated by particular livelihoods; poor and crowded living conditions; lack of infrastructure related to potable water, electricity, and road networks; residing on marginal land; and many other factors. Consequently, some of the households most vulnerable to climate change are families that rely on subsistence or small-scale agriculture and, thus, are uniquely dependent on rainfall and local weather conditions. However, urban families are also vulnerable to climate change. The urban poor often reside in low elevation coastal areas and lack access to adequate housing and infrastructure (Balk et al., 2009). They may also face higher temperatures because of lack of access to air conditioning, poor or inadequate employment conditions, and high levels of pollution, causing significant impacts on health (Bekkar et al., 2020; Hsu et al., 2021; Isen et al., 2017). Furthermore, within these rural and urban families, pregnant women and children may be especially vulnerable.

In this section, we describe the Demographic and Health Surveys, a key data source for evaluating health in low- (and some) middle-income countries, and other publicly available household surveys in developing countries that allow researchers to study the impacts of climate change on a wide range of outcomes, for a range of at-risk and vulnerable groups.

Demographic and Health Surveys

Since our work focuses on maternal and child health in low-income settings, a major source of socio-demographic data has been the DHS, which have been conducted in over 90 countries in Africa, Asia, Latin America, and the Caribbean (The DHS Program – Country List, n.d.). The DHS surveys are administered to a representative sample of the study population and provide detailed health and demographic information (see Table 3.2). Despite the DHS origins as a follow-up to World Fertility Surveys, the DHS is a good source of data on families and men. The DHS includes household, female, and male questionnaires; furthermore, some surveys and modules include wide age ranges (e.g., the Haiti 2016/7 DHS surveyed men between the ages of 15 and 64 years). With the DHS, researchers can study the impacts of climate change on fertility (Grace, 2017; Thiede et al., 2020), family planning, reproductive health, breastfeeding practices, child health/mortality (Geruso & Spears, 2018), nutrition (Grace et al., 2012), anemia, malaria (Grace et al., 2021; Kudamatsu et al., 2012), HIV/AIDS (Baker, 2020), and domestic violence (Cools et al., 2020; Epstein et al., 2020).

Spatial Identifiers and Scales in DHS

Most recent DHS surveys provide geocoded information for the primary sampling unit in the surveys (i.e., clusters). DHS clusters represent groups of 20–40 households (Fig. 3.1a). The cluster geocodes are assumed to be the geographic centroids of the communities. Consequently, in an urban setting where there is greater population density, they typically represent households in a small geographic area. In contrast, in sparsely populated rural areas, the geocode for a cluster can be located at the centroid of a large geographic area if households are pulled from different villages (Burgert et al., 2013; Mayala et al., 2018). The DHS randomly displaces the geocodes of the urban clusters up to 2 km and of rural clusters up to 5 km, selecting a small portion of the data for a 10 km shift. Beginning in 2009, displaced clusters remain within the country's second administrative level. This strategy to maintain confidentiality while using detailed, spatially referenced health survey data is also employed by the World Bank (with the Living Standard and Measurement Study - Integrated Surveys on Agriculture data) and the Gates Foundation's Performance Monitoring for Action data. In our research using the DHS clusters, we accommodate the random displacement by including a 10 km radius around each georeferenced cluster (Davenport et al., 2020; Grace et al., 2021; Perez-Heydrich et al., 2016). However, depending on the research question and climate data employed, alternative strategies for accommodating different spatial scales in the data may more accurately reflect community-level heterogeneity and be useful for uncovering associations in the data (Grace et al., 2019).

Table 3.2 Common household surveys in developing countries that can be linked with climatic data

Household surveys	Survey population	Finest spatial detail	Temporal details
Demographic and health surveys (DHS)	Household residents and visitors; women and men aged 15–49 years; children under the age of 5 years	Geocodes (lat/long) for enumeration areas (15–30 households)	Cross-sectional survey, reproductive histories, interview dates, birth dates
Malaria indicator survey (MIS)	Households; women aged 15–49 years; children under the age of 5 years	Geocodes (lat/long) for enumeration areas (15–30 households)	Cross-sectional survey, reproductive histories for 5 years prior to interview
Multiple indicator cluster surveys (MICS)	Household residents; women and men aged 15–49 years; children aged 5–17 years; children under the age of 5 years	Administrative unit level 1. Some countries such as Bangladesh have admin level 2 data. GPS may be available by requesting special access from the country's statistical office	Cross-sectional survey, reproductive histories, interview dates, birth dates
Performance monitoring for action (PMA)	Households; women aged 15–49 years	Geocodes (lat/long) for enumeration areas	Panel survey, annual data collection
World Bank living standard measurement study – Integrated surveys on agriculture (LSMS-ISA)	Households; communities	Geocodes (lat/long) for enumeration areas (roughly village scale)	Panel survey

Note: Descriptions are based on the characteristics of the latest survey rounds as of 2021

To reduce some of the challenges to users, a range of geospatial covariates (contextual variables), including climate variables, are now available through various platforms such as the DHS website, IPUMS DHS, and AReNA (IFPRI, 2020; IPUMS-DHS, n.d.). These data provide the user with pre-processed summaries of commonly employed (or user requested) variables (Mayala et al., 2018). The summaries are typically generated for the DHS sampling cluster (as opposed to the individual respondent), and, depending on the variable, a monthly, yearly, or 5-year summary measure such as average, maximum, or minimum is provided. However, this is a "user beware" situation as there exists an underlying spatial and temporal structure of the data that impacts the interpretation of the variables when used. Thus, users are strongly encouraged to read the metadata accompanying these datasets and consider the data-generating processes and the spatial and temporal aggregation of the variables to ensure that they are using appropriate measures to advance their research.

Temporal Identifiers in DHS and Cross-Sectional and Longitudinal Data

DHS surveys collect the interview date, biomarker collection date, and dates/length for key events in the life course of individuals: marriage date, respondent birth date, children's birth date, length of pregnancy in months, and duration of breastfeeding (ICF, 2018). Prior to 2015, DHS dates were in month-year format (century month code); thus, the day was missing. Starting with DHS version 7 (DHS-7), more detailed dates are available (century day code), which allows the analyst to report the exact day, month, and year of birth. The DHS-7 also includes length of pregnancy, which allows analysts to better estimate the conception date. When dates are not reported or are inconsistent, the DHS often imputes the missing dates (ICF, 2018); consequently, analysts should be aware of and have a strategy for dealing with imputed dates. Even when excluding imputed dates, there have been some concerns about the accuracy of dates in DHS (Larsen et al., 2019).

The DHS are repeated cross-sectional surveys which typically take place every 5 years. However, two elements of the DHS allow for a type of longitudinal analyses. The DHS contains detailed calendar data which allows analysts to obtain month-by-month 5-year retrospective data of contraceptive use and reproductive outcomes (i.e., pregnancies, births, and terminations; Davenport et al., 2020). Similarly, the retrospective birth histories record all the births a woman has had and therefore are not limited to the 5 years prior to the survey date. The reproductive histories collect information on the date of birth, survival status of the births, and date of child death, if applicable. Therefore one can create a month-by-month panel of child survival (Dorélien, 2015; Geruso & Spears, 2018). One caveat, which may introduce bias, is that these longitudinal measures are present only for women who are alive at the time of survey. Furthermore, the longitudinal data derived from the DHS in this way do not typically contain information on temporally varying measures of many dynamic factors related to health (e.g., menstruation, child health, abortion/conception, contraceptive demand, mobility, employment/income, etc.).

Linking Spatial Household Surveys with Climate Datasets

Another major source of spatial demographic and health data is UNICEF's *Multiple Indicator Cluster Surveys (MICS*; Khan & Hancioglu, 2019). These household surveys have been conducted in 118 countries and are very similar in content to the DHS (Hancioglu & Arnold, 2013). For instance, the MICS also provides information on child mortality and health, nutrition, education, and HIV/AIDs. Unlike the DHS, the MICS collects information on child labor and discipline and early child development. Another key difference between the MICS and DHS is that the reference period in MICS is shorter than in DHS. The MICs survey instrument collects information on birth weight for the last birth in the past 2 years, while DHS collects information on all births in the past 5 years. Another important difference is that the DHS collects information on children from only the biological mothers in the women's questionnaire, while the MICS collects information on children from either mothers or caregivers. Consequently, the MICS also contains information on orphans and foster children (Hancioglu & Arnold, 2013). Unlike the DHS, MICS does not provide geocoded clusters; the finest spatial identifier which can be linked to climate exposure, the administrative level one unit, is coarse.

Performance Monitoring for Action (PMA) surveys are household, female, and facility surveys that collect information on sexual and reproductive health and are currently administered in nine high-fertility countries in Asia and Africa (data is available for 11 countries; IPUMS PMA, n.d.). While not all data is spatially referenced, many survey rounds include geocodes similar to the DHS (i.e., enumeration cluster locations that are displaced to protect confidentiality of respondents). The PMA surveys have also collected data on the impacts of COVID-19, child nutrition, maternal and newborn health, migration histories, and healthcare service delivery context. However, the samples can vary in notable ways between countries and survey rounds as the particular questionnaires are designed to reflect country-specific interests. The first PMA survey took place in 2013; data collection takes place at least annually. Since 2019, most of the surveys have been modified to a panel design to follow the same women over time, providing longitudinal data to monitor temporal changes in exposures and outcomes among the study respondents (IPUMS PMA, n.d.).

The Malaria Indicator Surveys (MIS) are stand-alone cross-sectional surveys based on DHS and MICS and are available for about 30 countries. The scope of the MIS is narrower than for the DHS or MICS. The MIS contains information on household roster, assets, and use of insecticide-treated nets. The female questionnaire includes the reproductive birth histories and care-seeking information for children under age 5 years; it does not contain complete reproductive histories. The biomarker questionnaire records biomarkers of anemia and malaria infections for children between ages 6 and 60 months (The DHS Program – Malaria Indicators Survey [MIS], n.d.). One key feature of the MIS surveys is that they are timed to correspond with the malaria transmission season in the surveyed countries. As in the DHS, latitude and longitude coordinates are available for the MIS survey clusters.

World Bank's Living Standards and Measurement Study – Integrated Surveys on Agriculture (LSMS-ISA) Similar to the DHS, the World Bank's LSMS-ISA provides information on household and individual factors, including some anthropometric measurements (LSMS-ISA, n.d.). LSMS-ISA, however, specifically gathers information about consumption, expenditures, and household assets for a selection of households. The LSMS-ISA also provides additional information on agricultural production and livestock holdings for each household and gathers this information at two time periods: pre- and post-harvest period. Additionally, for some countries, the LSMS-ISA follows the same households over several survey rounds, creating a longitudinal record of household-level agricultural production and child anthropometrics. Notably, the LSMS-ISA does not contain detailed retrospective birth information and rarely includes information about migration or prior place of residence. The LSMS-ISA provides some environmental covariates (e.g., annual mean temperature, annual precipitation, land cover class, and others). As with DHS and PMA, the LSMS-ISA provides offset geocodes of enumeration areas.

Identifying the Exposure Population

When studying the impact of climate exposures on health and demographic outcomes of populations, it is important to keep the exposure population in mind (Grace et al., 2020; Hunter et al., 2021). In other words, it is critical to identify which individuals were exposed to an event (i.e., the numerator) based on the temporal and spatial overlap between the individual and the event of interest. Mobility, migration, and mortality may make this task particularly complicated. For example, if climate change induces migration, then survey data is being collected only among those remaining in the area, rather than from the entire affected community, leading to possible selection/sampling bias. The population that has not migrated (e.g., those included in the survey) may be different from the population prior to the exposure (Hunter et al., 2021). Similarly, the population may contain recent migrants who were not exposed to an event that occurred in a given place. Full migration histories (even including short-term and seasonal migrations) with details on the age at migration and the motivation for migration (or for staying after an event) are generally not collected in surveys focused on health and demographic characteristics. Additionally, if mortality increases because of a climate event, then mortality selection may occur leaving those who survived as a distinctly different population from the original exposure population.

Given the lack of detailed migration data in the standard health surveys, there are some strategies for avoiding possible exposure misclassification. One strategy is to consider the length of time at current residence as previously described. However, it is important to note that when we exclude migrants from the analytic population (because they may not have been exposed), we may be introducing other types of bias since they may have unique experiences that make them either more robust or vulnerable than their non-migrant counterparts. In this way, the non-migrants may be negatively (or positively) selected compared to the non-mobile population of the community. For mortality, considering different techniques to examine selection versus scarring hypotheses may be useful.

Multilevel (Contextual) Analyses Link Climate Change and Health

The structure of the DHS and similar social spatial household surveys described here (see Table 3.2) allow applied scientists to study socioeconomic, behavioral, and biological responses to climate change through multilevel (e.g., individual, household, local context) analyses, especially when survey data is combined with other sources of data (e.g., climate data). There is an established literature linking climate change to adverse health and well-being outcomes among women and children. The research, focused specifically on childbearing and infant/child health, often considers how climate change impacts health through direct or indirect determinants (Bakhtsiyarava et al., 2018; Bekkar et al., 2020; Eissler et al., 2019). This research documents how climate change-related factors such as hot temperatures adversely impact pregnancy outcomes and fetal growth, leading to pre-term birth, low birth weight, and impaired infant and childhood growth (Bakhtsiyarava & Grace, 2021; Randell et al., 2021; Shively, 2017). Other climate change-related factors, such as drought or seasonal growing conditions, are linked primarily through food insecurity but also through increased risk of dehydration and infectious disease transmission (e.g., because of reliance on unclean water; Kumar et al., 2016; Pinchoff et al., 2021). Using the existing literature as a guide, Fig. 3.3 summarizes

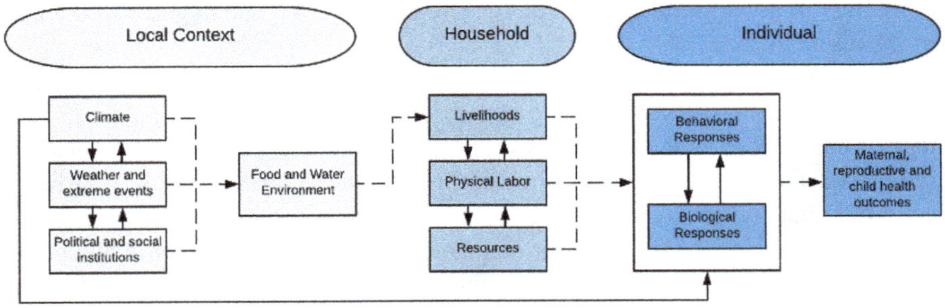

Fig. 3.3 Multilevel framework for linking maternal, reproductive, and child health outcomes with contextual factors

Note: The solid line demonstrates a direct link from climate to behavioral and biological responses; the dashed line indicates an indirect linkage. (Source: (Grace, 2017))

the different potential pathways linking climate exposures on the left (belonging to either the climate or weather/extreme events boxes in the graphic) and maternal, reproductive, and child health outcomes on the right (Grace, 2017). Direct linkages (solid line), such as heat waves, create stress on pregnant women and babies, leading to adverse pregnancy, birth, and infant/child health outcomes. Indirect linkages (dashed line) also cause adverse outcomes but operate through mechanisms such as disease, food insecurity, and inadequate nutrition.

Geographical factors such as level of urbanicity may cause climate exposures to have heterogeneous impacts on outcomes of interests. For example, the impact of heat waves on health outcomes is potentially worse in urban areas because of the heat-island effect where temperatures are likely to be hotter than in rural areas (Chakraborty et al., 2019; Hsu et al., 2021; Manoli et al., 2019), whereas drought impacts may be worse in rural areas where the food and water supply can be significantly impacted for extended periods of time, depending on the severity and duration of the drought (Cooper et al., 2019; Kumar et al., 2016). In both cases, improved infrastructure related to factors such as healthcare, air conditioning, fans, and resources to buy water or food at potentially elevated prices can reduce the adverse impacts of heat waves and droughts (Hsu et al., 2021; Shively, 2017).

Environmental Change: Impact on Maternal and Child Health

Putting these different approaches and data together has produced a diverse set of empirical papers that explore a range of research questions within the broader topic of climate change and health. While the approaches and data match much of what we have already described – analysts employ some type of individual-level or aggregated health dataset and link it to rainfall and/or temperature data – theoretical linkages, climate data used, and temporal and spatial data aggregation are notably different across existing studies. Here we briefly describe two recent projects that used similar data but considered different scopes and dimensions of children's health. We highlight the analytic approach used and how the findings build upon each other to highlight potential areas of vulnerability. We also note the limitations of the research.

In the first project (Davenport et al., 2020), we consider how pregnancy outcomes are impacted by in utero exposure to climate change factors (e.g., rainfall and temperature). To conduct this analysis, we used the calendar data from spatially referenced DHS data. We used all spatially referenced DHS data for 15 African countries that included the calendar and the length of time at current residence. Each pregnancy in the DHS (live birth outcomes as well as non-live birth outcomes) was linked in space and time to the count of hot days (days above a specific temperature threshold) and the average rainfall conditions within a trimester. Individual-level controls, such as educational attainment and age, were included in the analysis as well as household-level controls related to assets. Using a range of different statistical analysis methods that account for the changing exposure window, the results indicate that high temperatures increase the risk of non-live birth outcomes relative to live healthy births. Individual factors such as education were not able to reduce the overall risk that women faced. Thus, our findings indicate that when women experience more hot days during their pregnancies, their risk of miscarriage or stillbirth increases.

This research faced limitations. Importantly, we were not able to distinguish between induced and spontaneous abortions, and birth weight data (used to identify a healthy live birth) was not always available (and some colleagues questioned the quality of recalled birth weight data). Another challenge was the measurement of environmental variables. While we grouped all countries together (using country-level fixed effects), it was clear that this aggregation obscured some notable differences between countries. In particular, we noticed significant differences between West African countries as compared to those in the East.

To further investigate this, we developed an extension of the original study where we considered variation in only the birth weight of live births (using an indicator for whether the weight was recalled or noted on the birth record). We then focused the analysis specifically on a single country, Mali, in West Africa, the region where women seemed to face particular risks as determined by our prior analysis (Grace et al., 2021). Accounting for the fact that in Mali a good growing season may also create the perfect conditions for increased exposure to malaria, we developed a series of place-specific climate indicators that aimed to capture heat stress, food insecurity, and disease conditions related to malaria. While the results are somewhat similar to other related analyses – in general, exposure to more hot days reduces birth weight – the results (combined with the insight gained from the in utero exposure paper) also suggested that malarial conditions may in fact increase the risk of stillbirth or miscarriage among the frailest pregnancies and result in heavier infants at birth. Finally, the results suggest that improved agricultural seasons may reduce the severity of food insecurity during the hunger season, also increasing the weight of a child at birth.

The analyses revealed patterns consistent with our general understanding of climate-health but also helped us focus on a country where the patterns seemed particularly distinct. We then were able to use Mali's unique climate and landscape features to develop a series of climate indicators that were better able to capture the climate-health linkages than when developing a large multi-country analysis. In each case, the analyses relied on complex spatial and temporal data and careful aggregations of these data that reflected the distinct aspects of the particular research question.

Gaps and Opportunities in Studying the Impacts of Climate Change

Our DHS research, as well as that of many other scholars in the field, has focused on the impacts of climate change on women of reproductive age and on children because these are particularly vulnerable subpopulations but also because the large, spatially referenced, publicly available data (e.g., DHS) focuses on this subpopulation. Nevertheless, although we focus on causal pathways that act through the mother, because these are indirect measures of exposure, we may also be capturing effects due to paternal, familial, household, or community pathways.

A major limitation of the existing surveys and resulting datasets has been the lack of information on the elderly; individuals over age 65 years are rarely interviewed though the household rosters allow researchers to identify households with elderly residents. This is an important gap because the elderly may be especially vulnerable to climate change (Balbus & Malina, 2009). While the DHS contains a male questionnaire, the impact of climate change on adult men has also been largely unexplored. One exception is a recent analysis using both the female and male questionnaires that found evidence that increased temperature is positively correlated with male migration and sex market use in sub-Saharan Africa which has implications for HIV transmission (Baker, 2020). DHS studies on the effects of climate change have not focused on couples, although the survey contains variables that allow male and female responses of couples to be linked. The DHS surveys are not well suited for studying household adaptation to or coping strategies for climate exposures because of the cross-sectional nature of the surveys and limited data in the household questionnaires. The LSMS-ISA datasets are better suited to study household economic coping strategies since they include longitudinal datasets and contain information on agricultural activities/production and labor and non-labor market activities.

Moving forward, expanding spatially referenced individual-level data to include a much wider range of age groups and more directly address the ways that individuals, families, households, and communities manage resources in the face of climate change would dramatically expand scientific understanding and policy responses to climate change. Climate-health research has the potential to shape climate change mitigation policy and humanitarian aid interventions (as in the efforts of the USAID-funded www.fews.net effort). However, it is vital that social scientists who focus on individual-, family-, and community-level systems bring their expertise into these discussions. This type of research represents an innovative combination of disciplinary perspectives and therefore typically requires that social scientists and physical scientists build collaborations. Incorporating local, stakeholder knowledge can also significantly improve the quality and impact of the research (Grace & Mikal, 2019). Additionally, theoretical and applied analyses that consider multiple and cumulative impacts of climate events, or the interaction of climate and other events (such as conflict or economic instability), are also of great importance for advancing the field (Brown et al., 2021; Hill et al., 2019).

Acknowledgments We would like to thank Ashwini Shridhar, Divya Pandey, and Maya Luetke for their thoughtful comments on drafts of this chapter. The authors gratefully acknowledge support from the Minnesota Population Center (P2C HD041023) funded through a grant from the Eunice Kennedy Shriver National Institute for Child Health and Human Development (NICHD).

References

Baker, R. E. (2020). Climate change drives increase in modeled HIV prevalence. *Climatic Change, 163*(1), 237–252. https://doi.org/10.1007/s10584-020-02753-y

Baker, R. E., & Anttila-Hughes, J. (2020). Characterizing the contribution of high temperatures to child undernourishment in sub-Saharan Africa. *Scientific Reports, 10*, 18796. https://doi.org/10.1038/s41598-020-74942-9

Bakhtsiyarava, M., & Grace, K. (2021). Agricultural production diversity and child nutrition in Ethiopia. *Food Security*. https://doi.org/10.1007/s12571-021-01173-9

Bakhtsiyarava, M., Grace, K., & Nawrotzki, R. J. (2018). Climate, birth weight, and agricultural livelihoods in Kenya and Mali. *American Journal of Public Health, 108*(S2), S144–S150. https://doi.org/10.2105/AJPH.2017.304128

Balbus, J. M., & Malina, C. (2009). Identifying vulnerable subpopulations for climate change health effects in the United States. *Journal of Occupational and Environmental Medicine, 51*(1), 33–37. https://doi.org/10.1097/JOM.0b013e318193e12e

Balk, D., Montgomery, M. R., McGranahan, G., Kim, D., Mara, V., Todd, M., Buettner, T., & Dorélien, A. (2009). Mapping urban settlements and the risks of climate change in Africa, Asia and South America. *Population Dynamics and Climate Change, 80*. https://www.researchgate.net/publication/242472637_Mapping_Urban_Settlements_and_the_Risks_of_Climate_Change_in_Africa_Asia_and_South_America

Bekkar, B., Pacheco, S., Basu, R., & DeNicola, N. (2020). Association of air pollution and heat exposure with preterm birth, low birth weight, and stillbirth in the US: A systematic review. *JAMA Network Open, 3*(6), e208243. https://doi.org/10.1001/jamanetworkopen.2020.8243

Brown, M. E., Grace, K., Billing, T., & Backer, D. (2021). Considering climate and conflict conditions together to improve interventions that prevent child acute malnutrition. *The Lancet Planetary Health, 5*(9), e654–e658. https://doi.org/10.1016/S2542-5196(21)00197-2

Burgert, C. R., Colston, J., Roy, T., & Zachary, B. (2013). *Geographic displacement procedure and georeferenced data release policy for the Demographic and Health Surveys* (DHS spatial analysis reports no. 7). ICF International. https://dhsprogram.com/publications/publication-SAR7-Spatial-Analysis-Reports.cfm

Chakraborty, T., Hsu, A., Manya, D., & Sheriff, G. (2019). Disproportionately higher exposure to urban heat in lower-income neighborhoods: A multi-city perspective. *Environmental Research Letters, 14*(10), 105003. https://doi.org/10.1088/1748-9326/ab3b99

Cools, S., Flatø, M., & Kotsadam, A. (2020). Rainfall shocks and intimate partner violence in sub-Saharan Africa. *Journal of Peace Research, 57*(3), 377–390. https://doi.org/10.1177/0022343319880252

Cooper, M. W., Brown, M. E., Hochrainer-Stigler, S., Pflug, G., McCallum, I., Fritz, S., Silva, J., & Zvoleff, A. (2019). Mapping the effects of drought on child stunting. *Proceedings of the National Academy of Sciences, 116*(35), 17219–17224. https://doi.org/10.1073/pnas.1905228116

Davenport, F., Dorélien, A., & Grace, K. (2020). Investigating the linkages between pregnancy outcomes and climate in sub-Saharan Africa. *Population and Environment, 41*(4), 397–421. https://doi.org/10.1007/s11111-020-00342-w

Deschênes, O., Greenstone, M., & Guryan, J. (2009). Climate change and birth weight. *American Economic Review, 99*(2), 211–217. http://www.jstor.org/stable/25592401

Dorélien, A. (2015). Effects of birth month on child health and survival in sub-Saharan Africa. *Biodemography and Social Biology, 61*(2), 209–230. https://doi.org/10.1080/19485565.2015.1032399

Eissler, S., Thiede, B. C., & Strube, J. (2019). Climatic variability and changing reproductive goals in sub-Saharan Africa. *Global Environmental Change, 57*, 101912. https://doi.org/10.1016/j.gloenvcha.2019.03.011

Epstein, A., Bendavid, E., Nash, D., Charlebois, E. D., & Weiser, S. D. (2020). Drought and intimate partner violence towards women in 19 countries in sub-Saharan Africa during 2011-2018: A population-based study. *PLoS Medicine, 17*(3), e1003064. https://doi.org/10.1371/journal.pmed.1003064

Fussell, E., Sastry, N., & VanLandingham, M. (2010). Race, socioeconomic status, and return migration to New Orleans after hurricane Katrina. *Population and Environment, 31*(1–3), 20–42. https://doi.org/10.1007/s11111-009-0092-2

Geruso, M., & Spears, D. (2018). *Heat, humidity, and infant mortality in the developing world* (Working paper no. 24870; Working paper series). National Bureau of Economic Research. https://doi.org/10.3386/w24870.

Grace, K. (2017). Considering climate in studies of fertility and reproductive health in poor countries. *Nature Climate Change, 7*(7), 479–485. https://doi.org/10.1038/nclimate3318

Grace, K., & Mikal, J. (2019). Incorporating qualitative methodologies and fieldwork into large scale, quantitative analyses of climate health in low-income countries. *The Lancet Planetary Health, 3*(12), e496–e498. https://doi.org/10.1016/S2542-5196(19)30240-2

Grace, K., Billingsley, S., & Van Riper, D. (2020). Building an interdisciplinary framework to advance conceptual and technical aspects of population-environment research focused on women's and children's health. *Social Science & Medicine, 250*, 112857. https://doi.org/10.1016/j.socscimed.2020.112857

Grace, K., Davenport, F., Funk, C., & Lerner, A. M. (2012). Child malnutrition and climate in sub-Saharan Africa: An analysis of recent trends in Kenya. *Applied Geography, 35*(1), 405–413. https://doi.org/10.1016/j.apgeog.2012.06.017

Grace, K., Nagle, N. N., Burgert-Brucker, C. R., Rutzick, S., Van Riper, D. C., Dontamsetti, T., & Croft, T. (2019). Integrating environmental context into DHS analysis while protecting participant confidentiality: A new remote sensing method. *Population and Development Review, 45*(1), 197–218. https://doi.org/10.1111/padr.12222

Grace, K., Verdin, A., Dorélien, A., Davenport, F., Funk, C., & Husak, G. (2021). Exploring strategies for investigating the mechanisms linking climate and individual-level child health outcomes: An analysis of birth weight in Mali. *Demography, 58*(2), 499–526. https://doi.org/10.1215/00703370-8977484

Green, H., Bailey, J., Schwarz, L., Vanos, J., Ebi, K., & Benmarhnia, T. (2019). Impact of heat on mortality and morbidity in low and middle income countries: A review of the epidemiological evidence and considerations for future research. *Environmental Research, 171*, 80–91. https://doi.org/10.1016/j.envres.2019.01.010

Groen, J. A., & Polivka, A. E. (2010). Going home after hurricane Katrina: Determinants of return migration and changes in affected areas. *Demography, 47*(4), 821–844. https://doi.org/10.1007/BF03214587

Hancioglu, A., & Arnold, F. (2013). Measuring coverage in MNCH: Tracking progress in health for women and children using DHS and MICS household surveys. *PLoS Medicine, 10*(5), e1001391. https://doi.org/10.1371/journal.pmed.1001391

Hill, R., Skoufias, E., & Maher, B. (2019). *The chronology of a disaster: A review and assessment of the value of acting early on household welfare*. World Bank Group: Open Knowledge Repository. https://doi.org/10.1596/31721

Hsu, A., Sheriff, G., Chakraborty, T., & Manya, D. (2021). Disproportionate exposure to urban heat island intensity across major US cities. *Nature Communications, 12*(1), 2721. https://doi.org/10.1038/s41467-021-22799-5

Hunter, L. M., Koning, S., Fussell, E., King, B., Rishworth, A., Merdjanoff, A., Muttarak, R., Riosmena, F., Simon, D. H., Skop, E., & Van Den Hoek, J. (2021). Scales and sensitivities in climate vulnerability, displacement, and health. *Population and Environment, 43*(1), 61–81. https://doi.org/10.1007/s11111-021-00377-7

ICF. (2018). *Demographic and Health Surveys standard recode manual for DHS7*. ICF. Standard Recode Manual for DHS7 [DHSG4] (dhsprogram.com).

International Food Policy Research Institute (IFPRI). (2020). AReNA's DHS-GIS database. https://doi.org/10.7910/DVN/OQIPRW, Harvard Dataverse, V1, UNF:6:CCnbCvRUu7F/IAy2ut+whw== [fileUNF].

IPUMS-DHS. (n.d.). Retrieved September 12, 2021, from https://www.idhsdata.org/idhs/contextual_variables_download.shtml

IPUMS PMA. (n.d.). Retrieved September 12, 2021 from https://pma.ipums.org/pma/

Isen, A., Rossin-Slater, M., & Walker, R. (2017). Relationship between season of birth, temperature exposure, and later life wellbeing. *Proceedings of the National Academy of Sciences, 114*(51), 13447–13452. https://doi.org/10.1073/pnas.1702436114

Khan, S., & Hancioglu, A. (2019). Multiple indicator cluster surveys: Delivering robust data on children and women across the globe. *Studies in Family Planning, 50*(3), 279–286. https://doi.org/10.1111/sifp.12103

Kudamatsu, M., Persson, T., & Strömberg, D. (2012). *Weather and infant mortality in Africa* (SSRN Scholarly Paper ID 2210191). Social Science Research Network. https://papers.ssrn.com/abstract=2210191

Kumar, S., Molitor, R., & Vollmer, S. (2016). Drought and early child health in rural India. *Population and Development Review, 42*(1), 53–68. https://doi.org/10.1111/j.1728-4457.2016.00107.x

Larsen, A. F., Headey, D., & Masters, W. A. (2019). Misreporting month of birth: Diagnosis and implications for research on nutrition and early childhood in developing countries. *Demography, 56*(2), 707–728. https://doi.org/10.1007/s13524-018-0753-9

Living standards measurement study – Integrated studies on agriculture (LSMS-ISA). (n.d.). [Text/HTML]. World Bank. Retrieved September 13, 2021, from https://www.worldbank.org/en/programs/lsms/initiatives/lsms-ISA

LoPalo, M. (2020). *Temperature, worker productivity, and adaptation: Evidence from survey data production* (SSRN Scholarly Paper ID 3711044). Social Science Research Network. https://doi.org/10.2139/ssrn.3711044.

Manoli, G., Fatichi, S., Schläpfer, M., Yu, K., Crowther, T. W., Meili, N., Burlando, P., Katul, G. G., & Bou-Zeid, E. (2019). Magnitude of urban heat islands largely explained by climate and population. *Nature, 573*(7772), 55–60. https://doi.org/10.1038/s41586-019-1512-9

Mayala, B., Fish, T. D., Eitelberg, D., & Dontamsetti, T. (2018). *The DHS Program geospatial covariate datasets manual* (2nd ed.) p. 50. ICF. DHS Covariates Extract Data Description 2.pdf (dhsprogram.com).

McMichael, A. J. (2013). Globalization, climate change, and human health. *New England Journal of Medicine, 368*(14), 1335–1343. https://doi.org/10.1056/NEJMra1109341

McMichael, A. J., Woodruff, R. E., & Hales, S. (2006). Climate change and human health: Present and future risks. *The Lancet, 367*(9513), 859–869. https://doi.org/10.1016/s0140-6736(06)68079-3

Meehl, G. A., & Tebaldi, C. (2004). More intense, more frequent, and longer lasting heat waves in the 21st century. *Science, 305*(5686), 994–997. https://doi.org/10.1126/science.1098704

Parsons, L. (2019). Structuring the emotional landscape of climate change migration: Towards climate mobilities in geography. *Progress in Human Geography, 43*(4), 670–690. https://doi.org/10.1177/0309132518781011

Perez-Heydrich, C., Warren, J. L., Burgert, C. R., & Emch, M. E. (2016). Influence of demographic and health survey point displacements on raster-based analyses. *Spatial Demography, 4*(2), 135–153. https://doi.org/10.1007/s40980-015-0013-1

Phalkey, R. K., Aranda-Jan, C., Marx, S., Höfle, B., & Sauerborn, R. (2015). Systematic review of current efforts to quantify the impacts of climate change on undernutrition. *Proceedings of the National Academy of Sciences, 112*(33), E4522–E4529. https://doi.org/10.1073/pnas.1409769112

Pinchoff, J., Turner, W., & Grace, K. (2021). The association between agricultural conditions and multiple dimensions of undernutrition in children 6-23 months of age in Burkina Faso. *Environmental Research Communications, 3*(6), 065004. https://doi.org/10.1088/2515-7620/ac07f5

Randall, S., Coast, E., Antoine, P., Compaore, N., Dial, F.-B., Fanghanel, A., Gning, S. B., Thiombiano, B. G., Golaz, V., & Wandera, S. O. (2015). UN census "households" and local interpretations in Africa since independence. *SAGE Open, 5*(2). https://doi.org/10.1177/2158244015589353

Randell, H., Grace, K., & Bakhtsiyarava, M. (2021). Climatic conditions and infant care: Implications for child nutrition in rural Ethiopia. *Population and Environment, 42*(4), 524–552. https://doi.org/10.1007/s11111-020-00373-3

Rao, N., Lawson, E. T., Raditloaneng, W. N., Solomon, D., & Angula, M. N. (2019). Gendered vulnerabilities to climate change: Insights from the semi-arid regions of Africa and Asia. *Climate and Development, 11*(1), 14–26. https://doi.org/10.1080/17565529.2017.1372266

Rao, N., Singh, C., Solomon, D., Camfield, L., Sidiki, R., Angula, M., Poonacha, P., Sidibé, A., & Lawson, E. T. (2020). Managing risk, changing aspirations and household dynamics: Implications for wellbeing and adaptation in semi-arid Africa and India. *World Development, 125*, 104667. https://doi.org/10.1016/j.worlddev.2019.104667

Segnon, A. C., Totin, E., Zougmoré, R. B., Lokossou, J. C., Thompson-Hall, M., Ofori, B. O., Achigan-Dako, E. G., & Gordon, C. (2021). Differential household vulnerability to climatic and non-climatic stressors in semi-arid areas of Mali, West Africa. *Climate and Development, 13*(8), 697–712. https://doi.org/10.1080/17565529.2020.1855097

Shively, G. E. (2017). Infrastructure mitigates the sensitivity of child growth to local agriculture and rainfall in Nepal and Uganda. *Proceedings of the National Academy of Sciences, 114*(5), 903–908. https://doi.org/10.1073/pnas.1524482114

Shukla, S., Husak, G., Turner, W., Davenport, F., Funk, C., Harrison, L., & Krell, N. (2021). A slow rainy season onset is a reliable harbinger of drought in most food insecure regions in sub-Saharan Africa. *PLoS One, 16*(1), e0242883. https://doi.org/10.1371/journal.pone.0242883

Sohnesen, T. P. (2020). Two sides to same drought: Measurement and impact of Ethiopia's 2015 historical drought. *Economics of Disasters and Climate Change, 4*(1), 83–101. https://doi.org/10.1007/s41885-019-00048-w

The DHS Program–Country List. (n.d.). Retrieved September 12, 2021, from https://dhsprogram.com/Countries/Country-List.cfm

The DHS Program–Malaria Indicators Survey (MIS). (n.d.). Retrieved September 12, 2021, from https://dhsprogram.com/methodology/survey-types/mis.cfm

Thiede, B. C., & Strube, J. (2020). Climate variability and child nutrition: Findings from sub-Saharan Africa. *Global Environmental Change, 65*, 102192. https://doi.org/10.1016/j.gloenvcha.2020.102192

Thiede, B. C., Chen, J., Mueller, V., Jia, Y., & Hultquist, C. (2020). *It's raining babies? Flooding and fertility choices in Bangladesh* (No. cz482). *Center for Open Science*. https://doi.org/10.31235/osf.io%2Fcz482.

Watts, N., Adger, W. N., Agnolucci, P., Blackstock, J., Byass, P., Cai, W., Chaytor, S., Colbourn, T., Collins, M., Cooper, A., Cox, P. M., Depledge, J., Drummond, P., Ekins, P., Galaz, V., Grace, D., Graham, H., Grubb, M., Haines, A., et al. (2015). Health and climate change: Policy responses to protect public health. *The Lancet, 386*(10006), 1861–1914. https://doi.org/10.1016/S0140-6736(15)60854-6

Xu, Z., FitzGerald, G., Guo, Y., Jalaludin, B., & Tong, S. (2016). Impact of heatwave on mortality under different heatwave definitions: A systematic review and meta-analysis. *Environment International, 89–90*, 193–203. https://doi.org/10.1016/j.envint.2016.02.007

Chapter 4
Family Well-Being in the Context of Environmental Migration

Amanda R. Carrico

The earth has warmed by approximately 1.0° Celsius in the past century, making the last several decades among the hottest in modern human history (IPCC, 2018). The extent of future warming is yet to be determined, though, even with rapid and aggressive mitigation, further warming is inevitable. As such, the question of how climate change is influencing human behavior and social systems has emerged as a central focus in contemporary social science. The unit of analysis for such inquiries is wide ranging, spanning questions about macro-level shifts in political and economic systems (Dell et al., 2008; Fiorino, 2018) to those emphasizing individual perceptions and decision-making (Grothmann & Patt, 2005; Truelove et al., 2015). Work that centralizes the family as the unit of analysis in climate change research is surprisingly rare, despite the primacy of families in critical decisions involving mobility, livelihoods, and the use and allocation of resources.

In this chapter, I seek to forge a connection between research on impacts of and responses to climate change within social systems and the field of family studies. In doing so, I focus specifically on the topic of migration. As concerns over climate change grow, an emerging narrative has emphasized the potential for large-scale out-migration and displacement, with projections ranging from several million to over 1 billion persons leaving their communities of origin in the coming century (Baird et al., 2007; Gemenne, 2011; Jacobson, 1988; Myers, 2002). In reality, the relationship between climate change and mobility is complex, and predictions are highly speculative (Gemenne, 2011; Lilleør & Van den Broeck, 2011; Tacoli, 2009). Although forced displacement does occur, carrying with it significant consequences, migration is increasingly understood as a multifaceted phenomenon that often functions as an adaptation strategy utilized by families alongside of other coping

A. R. Carrico (✉)
University of Colorado, Boulder, CO, USA
e-mail: amanda.carrico@colorado.edu

S. E. Ortiz et al. (eds.), *Environmental Impacts on Families*, National Symposium on Family Issues 12, https://doi.org/10.1007/978-3-031-22649-6_4

responses. Migration may be employed proactively to mitigate risk or reactively to manage the impacts of climate shocks. Environmental stress sometimes inhibits migration (Black et al., 2011; Nawrotzki & DeWaard, 2018), thereby constraining access to what has long been a pivotal strategy for families to access economic opportunity. When migration is pursued, it is often negotiated within families, is shaped by the human and social capital held within a household, and is conditioned by the cultural norms and practices within a place. Moreover, when families choose to send a migrant, the decision can have profound and sometimes countervailing impacts on the dynamics and well-being of other household members.

In the sections that follow, I review literature that examines the relationship between climate change and migration. I focus specifically on migrant trips that are pursued voluntarily, rather than instances of forced migration that are expertly discussed elsewhere (e.g., Berchin et al., 2017; McAdam, 2010; Wilmsen & Webber, 2015). I first discuss how the term climate migration has been defined and measured, with an emphasis on quantitative approaches. I follow this with a brief review of key findings regarding when, why, and how climate change influences migration. The question of how climate change affects the selectivity of migrants—effectively pushing those historically less likely to move into the migrant stream, or nudging others out—has received some recent attention (Hunter & Simon, 2017; Nawrotzki et al., 2015) and has particular relevance to the field of family studies. I seek to add to this discussion using data from the environmentally vulnerable and highly mobile context of coastal Bangladesh to investigate: Who are the environmental migrants and what might this mean for families and their well-being? I conclude with a discussion of insights for future efforts to understand climate change adaptation through the lens of family studies.

Theoretical Orientations and the Significance of Families

Research investigating the links between climate change and migration has proliferated in recent years, stimulating a series of methodological and conceptual advances that underpin current understanding in the field. The New Economics of Labor Migration (NELM) framework (Massey et al., 1999; Stark & Bloom, 1985) has emerged as a central organizing framework in recent decades. NELM marked a significant departure from prior theoretical orientations by shifting emphasis to the household as the unit of analysis in migration decision-making, rather than individual labor migrants. This approach understands migration as a household-level decision that is negotiated within families and used as a strategy to diversify risk. By sending a migrant, a family is investing in a source of income that is exposed to a different set of risks than income streams in origins. Through geographic dispersal and livelihood diversification across family members, households can reduce the vulnerability of their livelihood portfolios. Within the context of environmental migration, this approach understands migration as a critical pathway to livelihood adaptation rather than simply an individualized economic choice or a result of displacement. The NELM framework is supported by the finding that rural-to-urban migrants in low-income nations often maintain strong ties with their origins, contributing remittances to support other family members at home (Angelsen et al., 2020; Bohra-Mishra, 2013; Ngoma & Ismail, 2013).

Alternatively, it is common for scholars in the field to draw on theories that emphasize processes of adaptation and resilience. Here, migration is understood as one adaptation response among many that a household may pursue in the face of environmental stress (Barnett & Adger, 2007; Black et al., 2011). The Sustainable Livelihoods Framework (SLF) theorizes that livelihood resilience is grounded in access to multiple forms of assets or *capitals* (Bebbington, 1999). The SLF recognizes a range of capitals beyond simple economic resources, including human, social, physical, and natural capital. Livelihood activities contribute to the accumulation of capital resources, but also depend on the availability of those resources. Like NELM, the SLF considers the household to be the most relevant unit of analysis and seeks to understand the portfolio of assets held by a household, but SLF also emphasizes capital that accumulates (or not) within communities. With respect to migration decisions, the SLF points to the significance of human and social capital in offsetting economic costs or constraints. In other words, a family must have human capital in the form of individuals with the physical capacity and social opportunity to leave the origin to find work elsewhere. Likewise, social ties to other migrants in the destination are critical to facilitating access to transportation, employment, and housing. Those human and social resources alter the costs and benefits associated with various livelihood adaptations and are informative in understanding which families are likely to pursue migration relative to other forms of adaptation.

Defining and Operationalizing Climate Migration

Despite recent calls from world leaders for more efforts to protect the well-being of those pushed out and displaced by climate change (e.g., Biden Jr, 2021; IOM, 2021), no legal definition of *climate migration* exists (Chazalnoel & Ionesco, 2016; Dun & Gemenne, 2008). The International Organization for Migration (IOM) describes a broader class of *environmental migrants* as:

> A person or group(s) of persons who, predominantly for reasons of sudden or progressive changes in the environment that adversely affect their lives or living conditions, are forced to leave their places of habitual residence, or choose to do so, either temporarily or permanently, and who move within or outside their country of origin or habitual residence. (IOM, 2019)

IOM then defines *climate migration* as a subcategory of environmental migration where the environmental change is due to climate change. These definitions acknowledge that climate change impacts on migration are complex and difficult to estimate. Climate migration may be forced or pursued voluntarily as a form of adaptation, can involve domestic or international moves, and is often motivated by multiple interacting causal factors. Researchers in this field have similarly grappled with how to empirically define the term. Recognizing that complex and dynamic forces contribute to migration decisions, our ability to pinpoint a causal effect of climate change is especially challenging (Black et al., 2011; Kaczan & Orgill-Meyer, 2020; Renaud et al., 2011). When asked to explain the reason for migrating, most families cite economic motivations. For example, in my own experience conducting surveys in a region of Bangladesh that is highly exposed to environmental hazards, less than 1% of migrants (from over 4000 households) cite environmental conditions as the reason for their move. Yet, the economic conditions that lead individuals to migrate for work are often driven by environmental stress, especially for those in resource-dependent sectors (Hugo, 1996). Moreover, the causal link between climate impacts and decisions to migrate may be delayed or contingent upon an accumulation of impacts over time.

In response to the methodological challenges described above, most quantitative work in this field has inferred the effects of climate impacts by estimating proportional differences in rates of migration from a given place as a function of spatial or temporal variation in climate conditions. Data are most often generated from surveys[1] that record information about household members, including those who have left as reported by family who remained behind. Information is collected about past or ongoing migrant trips alongside of other sociodemographic characteristics (e.g., year of birth, year of marriage, livelihood histories). Starting with a household member's year of birth, these data are then used to construct a person-period dataset in which each case represents a "period" in a household member's life. Most commonly, the period refers to years; however, it may also reflect shorter or longer segments of time (e.g., months or decades). Variables representing whether a person made a migrant trip in a given period (or not) are added to the file, as are variables containing other fixed or time-varying information about the individual (e.g., gender, education, marital status), household (e.g., size, livelihoods, number of dependents), and community (e.g., infrastructure, density of migrants). Together, these data generate a rich dataset describing the life courses of individuals embedded within the context of families and communities (Fussell et al., 2014).

By integrating information about exposure to environmental shocks such as a heat wave, drought, or crop failure, we can analytically examine the relationship between climate shocks and the risk of migrating in the same or subsequent years (see Chap. 3, Dorélien & Grace, for a more detailed discussion of considerations when integrating climate data with sociodemographic information). This is most often done using hazards models that estimate whether rates of migration proportionately increase or fall during periods of exposure to environmental stressors (or not), controlling for other factors that might affect those decisions (e.g., Carrico &

[1]A full discussion of the range of methods used in this literature is beyond the scope of this chapter. For a more complete review of methods used in this field, including the use of census data, national registries, surveillance methods, agent-based or dynamic models, and qualitative approaches, please see, e.g., Eklund et al., 2016; Fussell et al., 2014; Hoffmann et al., 2021; Piguet, 2010.

Donato, 2019; Fussell et al., 2014; Gray & Mueller, 2012b; Sedova & Kalkuhl, 2020). Environmental variables may reflect not only exposure but also the intensity of events, and hierarchical models may be used to consider how relationships between climate and migration vary across context. Likewise, longitudinal surveys or surveillance approaches can improve upon cross-sectional data by capturing changes in individual, household, and community conditions that cannot be easily recalled and by accommodating prospective analysis.

It is important to note that, despite the ubiquitous use of the term, analyses such as those described previously do not capture climate change per se. *Climate change* refers to a change in the state of the climate, often assessed through analyses of the central tendency or variability of environmental indicators over an extended period of time (i.e., decades or longer). Multi-decadal data can reveal significant trends indicating that climate properties have changed, and models are often used to forecast future trends. However, most individual environmental events such as a heat wave or drought cannot be tied directly to climate change. This challenge of linking discrete outcomes to causal mechanisms that are probabilistic in nature is, of course, not unique to climate change research. As a result, climate and migration scholars study mobility in response to extreme weather, disasters, or other events with the understanding that they characterize the types of impacts that have or are expected to grow more frequent and intense due to anthropogenic climate change (Hoffmann et al., 2021; Kaczan & Orgill-Meyer, 2020). Therefore, specific interpretations referring to "climate migration" should be treated carefully.

Although comparatively less common (Zickgraf, 2021), several scholars have also attempted to capture exposure to slow-onset or cumulative environmental change that better reflects the technical definition of climate change. For example, Call and Gray (2020) estimated effects of moving average temperature and rainfall anomalies to predict different forms of migration (e.g., labor, non-labor), but did so separately over 12- and 120-month periods to differentiate effects associated with short-term vs longer-term exposures. Other similar approaches have considered whether the rate or severity of meteorological anomalies over an extended period of time predicts migration at a subsequent stage (e.g., Thiede et al., 2016). Alternatively, several studies have examined relationships between perceived long-term climate change and self-reported migration drawn from semi-structured interviews or survey data (e.g., Panda, 2017).

Empirical Studies: Impact of Environmental Stress on Migration

On balance, scholarship in this field has generated strong evidence that environmental stresses associated with climate change are related to rates and patterns of migration (Hoffmann et al., 2020). Generally, heat stress has emerged as one of the most consistent and important predictors. For example, exposure to prolonged heat waves or above average temperatures has been associated with an increase in outmigration in Indonesia (Bohra-Mishra et al., 2014), sub-Saharan Africa (Dillon et al., 2011), and South Asia (Call et al., 2017; Carrico & Donato, 2019; Mueller et al., 2014). The effect of precipitation has been demonstrated less consistently, but several studies reveal that drought or rainfall deficits are also associated with an overall increase in migration, especially in already arid regions (Hoffmann et al., 2020; Nawrotzki et al., 2015; Nawrotzki & DeWaard, 2016). The vast majority of work has focused on meteorological conditions including temperature and rainfall; however, other studies also reveal significant increases in migration in response to salinity (Chen & Mueller, 2018), cyclones (Mallick & Vogt, 2012; Saha, 2017), and flooding (Robalino et al., 2015).

Importantly, several studies also find evidence that climate shocks are associated with a decline in migration (Chen et al., 2017; Gray & Mueller, 2012b), which may suggest the presence of "trapped populations" or the possibility that households are delaying or foregoing planned trips during adverse conditions (Nawrotzki & DeWaard, 2018). Exactly when and why environmental shocks result in lower vs higher rates of migration is poorly understood and a gap to be filled by future research. Several findings point to the possibility that conditions that compromise transportation infrastructure, such as floods, may disrupt migration. Likewise, income shocks may deplete the resources needed to execute a trip, especially for the very poor or those living in isolated regions.

Other work reveals significant trends in the characteristics of migrant trips that are associated with environmental events. A common conclusion is that climate shocks are more often associated with short distance and short duration moves, including more domestic and seasonal trips (Gray & Mueller, 2012a; Henry et al., 2004; Hunter et al., 2013). However, relatively few studies report findings for internal moves, and recent work suggests that international trips may be equally as prevalent as internal moves (Hoffmann et al., 2020). Several studies also find that outmigration following a climate shock is more prevalent in low-income nations and in regions that are dependent on agriculture, which has led to the theory that disruption to resource-dependent livelihoods is a primary mechanism by which climate shocks lead to migration (Call et al., 2017; Carrico & Donato, 2019; Nawrotzki & Bakhtsiyarava, 2017).

An important observation within this literature is the extent to which findings vary by socioeconomic and environmental context. In response to strong evidence suggesting that climate change impacts are context-dependent, scholars have argued for a shift in focus that "[let's go] of ambitions for broadly applicable theory" and attempts to make far-reaching generalizations (Hunter, 2018, p. 3). There has been a renewed emphasis on understanding the processes of migration *within* the social, political, economic, and environmental dimensions (i.e., context) of a particular place. This includes closer examination of the characteristics of individuals and households that interact with climate to influence mobility.

Who Are the Environmental Migrants? A Case Study from Bangladesh

It is well-known that migration selects on social, demographic, and health characteristics—resulting in a community of migrants who tend to be younger, be healthier, be more educated, and hold fewer direct attachments to the origin household (e.g., via children or landholdings) than their non-migrant peers. In other words, migration selects for individuals who are better equipped for the complexity and physical demands of executing a migrant trip (Chiswick et al., 2008; Palloni & Morenoff, 2006). The high-risk nature of migration also deters those with more to lose from a failed trip (Lindstrom & López Ramírez, 2010). However, as we have observed an accumulation of evidence suggesting that climate variability contributes to migration, it also stands to reason that these pressures might simultaneously alter *who* migrates. Most research in the climate migration literature has focused on estimating the relationship between environmental shocks and migration. Yet, more work is needed to understand what are the characteristics of those who migrate in these circumstances. Unraveling the answer to this question has special importance for predicting the implications of climate change for families.

In this section, I draw on data from the Bangladesh Environment and Migration Survey (BEMS) to better understand the characteristics of migrants and migrant trips in years affected (or not) by an environmental shock. Bangladesh is widely regarded as one of the most environmentally vulnerable regions of the world. Approximately 70% of Bangladeshis live in a low-lying river basin exposed to flooding, erosion, and waterlogging. The nation's agricultural sector—which accounts for over 12% of GDP and employs 40% of the population—is especially vulnerable (Basak et al., 2010; Rahman et al., 2011). Migration is an important coping strategy used by rural households to access economic opportunity but also to manage the impacts of environmental shocks (Carrico & Donato, 2019; Donato et al., 2016; Gray & Mueller, 2012a; Mallick & Etzold, 2015). As a consequence, the number of Bangladeshis living in urban areas is expected to triple this century, with most urban growth coming from migration into informal urban settlements (Streatfield & Karar, 2008). Remittances sent from international destinations have been an important engine of economic growth. In 2019, personal remittances were over 6% of the nation's gross domestic product (GDP), placing Bangladesh in the top ten nations worldwide in total remittances received (World Bank, 2020). Remittances contribute to increases in health expenditures, improved sanitation, and improved food security in rural communities (Moniruzzaman, 2020; Sikder & Higgins, 2017).

In 2014, my colleagues and I implemented the BEMS in an initial sample of 1700 households in 9 communities in southwestern Bangladesh (see, e.g., Carrico & Donato, 2019; Donato et al., 2016). In 2019, we expanded the sample to include 20 new survey sites (n = 4001 households) in a wider geographic region (see Fig. 4.1). In the analysis presented here, I include data from this second phase only (BEMS-2). The BEMS draws on the methods used by the Mexican Migration Project to collect detailed social, economic, and demographic data from randomly

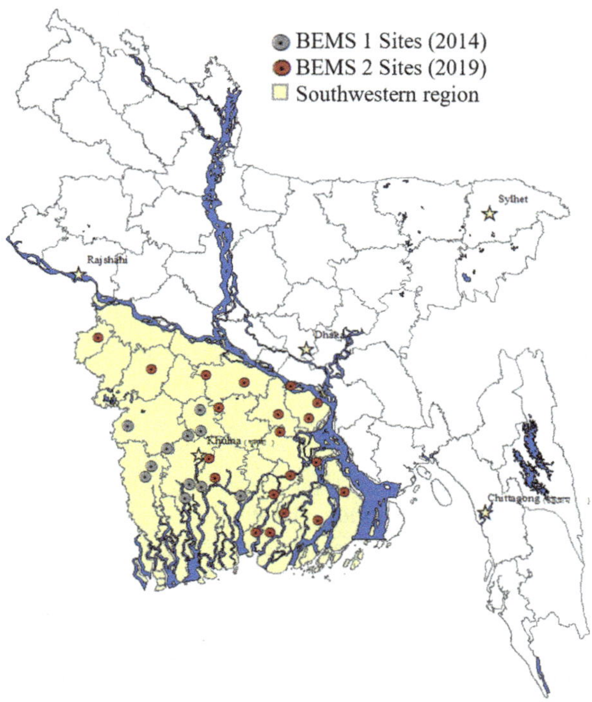

Fig. 4.1 Map of survey sites. Data from the BEMS-2, implemented in 2019, are included in this analysis

selected households. Communities in the BEMS refer to *mouzas*, small administrative districts typically consisting of 1–5 villages and a few hundred to a few thousand households. The 20 mouzas included in the BEMS-2 were selected from a sampling frame of all mouzas in the region after stratifying by district and socioeconomic status. To select households for inclusion in the survey, we first conducted a census of all households in each mouza and then randomly selected 200 per site.[2] As is common in this part of the world, we achieved a very high completion rate (95%).

Selected households were interviewed by trained local enumerators. The BEMS collects detailed migration histories from household heads, spouses, and other household members including years, destinations, and livelihood activities in the destination. If a household member was away at the time of the survey, we collected as much information as possible from those present. We also captured livelihood histories of heads and spouses, as well as self-reported health, environmental conditions, and demographic information about all household members. To mitigate recall error, we used several best practices in retrospective data collection, including cross-checking dates with the years of well-known community events and dates on government-issued identity cards (e.g., Auriat, 1991; Freedman et al., 1988; Smith & Thomas, 2003).

[2] If a community had fewer than 200 households, all were included in the survey, and the remaining sample was reallocated to more populated communities.

Defining Migrant Trips Taken During Periods of Disaster Exposure

The BEMS-2 recorded information about 8739 trips taken by 6078 adult (age 15+) household members in the BEMS sample. Although these trips spanned the period from 1944 to 2019, to mitigate recall bias and narrow the influence of sociopolitical changes over this time period, I constrain this analysis to the 6489 trips taken from 2000 to 2019. During this window of time, the southwestern region of Bangladesh experienced several significant environmental disasters. Most notable of these were Cyclones Sidr and Aila. Sidr made landfall on November 11, 2007, and affected more than nine million people (CRED, 2021). This included 4234 deaths and over 55,000 injured. Cyclone Aila made landfall less than 2 years later on May 25, 2009. It affected 3.4 million people and left 190 dead and 7103 injured. The much weaker Cyclone Mahasen made landfall in May of 2013, affecting 1.5 million persons and causing 17 recorded deaths. These events allow us to compare the characteristics of migrants and trips across years with and without significant environmental shocks.

To determine whether a migrant trip fell within a period impacted by a disaster, I integrated data from the BEMS that captured information about natural disasters directly experienced by the household. For each disaster reported, we recorded the year of impact, type and name of event (if applicable), and how the household was affected. Names and years of reported disasters were used to identify and resolve recall errors typically related to the year of the event. I considered a household to have been "exposed" if they reported impacts in the form of property loss or damage, lost job/income, or the injury/death of a household member. Ninety-five percent of all disasters reported within the period from 2000 to 2019 corresponded to one of the three cyclones mentioned here, though they were described in various terms (e.g., cyclone, flooding, storm surge, tidal surge). Although the remaining disasters were not easily identifiable, I retained these events because they are a small fraction of reported disasters and may reflect localized shocks stemming from riverine flooding, isolated storms, or embankment failures.

I used the information described above to classify whether a trip occurred during or after an environmental disaster. Recognizing that environmentally induced migration often follows several months or longer after a shock, I define the period of exposure as the year of or year after the household reported experiencing disaster impacts. With these definitions, 579 trips (8.9%) occurred during periods of disaster exposure and 5910 (91%) were taken during unexposed years.

Analyses and Findings

Table 4.1 presents statistics summarizing the characteristics of migrants and migrant trips. These statistics are presented for the full sample of trips but also broken out by whether the trip was taken during a period exposed to disaster impacts or not. All

Table 4.1 Migrant and trip characteristics across years exposed or not exposed to disasters

	All trips (6489)	Unaffected years (5910)		Disaster-affected years (579)		
	Mean	Mean	95% CI	Mean	95% CI	*p-value*
(A) Migrant demographics						
Male	0.72	0.72	0.70–0.73	0.74	0.69–0.78	0.321
Age	29.46	29.26	28.83–29.70	31.52	30.06–32.97	0.003
Married	0.66	0.65	0.64–0.66	0.75	0.71–0.80	0.000
# of children <14 in household	1.31	1.27	1.22–1.31	1.73	1.56–1.90	0.000
(B) Migrant socio-economic characteristics						
Literate	0.78	0.78	0.76–0.79	0.77	0.73–0.82	0.783
Education—Above primary	0.67	0.67	0.65–0.68	0.66	0.61–0.71	0.786
Education—Above secondary	0.35	0.35	0.33–0.37	0.32	0.27–0.38	0.324
Head in agriculture[a]	0.12	0.11	0.10–0.13	0.16	0.11–0.22	0.065
Head in unskilled work[a]	0.13	0.13	0.12–0.14	0.15	0.10–0.19	0.526
Head in business/prof[a]	0.28	0.27	0.26–0.29	0.34	0.28–0.41	0.021
(C) Health						
Height[b]	159.64	159.66	159.36–159.96	159.45	158.61–160.28	0.628
Self-rated healthy, before first trip[b]	0.75	0.75	0.72–0.77	0.74	0.66–0.82	0.906
Self-rated healthy, marriage[c]	0.78	0.77	0.75–0.80	0.84	0.78–0.91	0.088
(D) Trip characteristics						
Destination is foreign	0.17	0.17	0.16–0.18	0.17	0.13–0.21	0.930
Destination is Dhaka	0.40	0.41	0.39–0.42	0.36	0.30–0.41	0.122
Has migrant relatives	0.62	0.62	0.60–0.64	0.66	0.60–0.71	0.259
First trip	0.58	0.58	0.57–0.60	0.56	0.50–0.62	0.413
Duration >1 year	0.71	0.71	0.71–0.72	0.70	0.65–0.74	0.514
Duration >5 years	0.49	0.48	0.46–0.50	0.52	0.47–0.58	0.157
Livelihood sector						
Agriculture	0.01	0.01	0.01–0.01	0.02	0.00–0.05	0.165
Unskilled labor	0.06	0.06	0.05–0.07	0.07	0.05–0.10	0.384
Professional/business	0.23	0.23	0.22–0.24	0.23	0.18–0.28	0.997

[a]Reflects conditions in the year before the trip

[b]Height and self-rated health were collected from heads and spouses only. The sample size for these variables is unaffected years: $n = 2351$, affected years: $n = 207$

[c]Self-rated health at marriage was collected from heads and spouses only, and cases were dropped if the respondent was unmarried at the time of trip. The sample size for this variable is unaffected years: $n = 2260$, affected years: $n = 238$

estimates control for year and community fixed effects and cluster the standard errors at the household level to capture non-independence when multiple trips are associated with the same household.

Section A in Table 4.1 describes demographic characteristics of migrants. Like others, we find that migrants are disproportionately male (72%) and young with an average age of just under 30 years. There are some noticeable differences, though, between the two groups. During disaster-affected years, a significantly higher proportion of migrants are married (75% vs 66%) and come from households with more dependent children (1.73 vs 1.27), relative to unaffected years. Disaster-affected migrants also tend to be older by an average of about 2 years (31.52 vs 29.26). Together, these patterns suggest that disaster-affected migrants may carry relatively more household responsibilities and that environmental stress might motivate moves that would otherwise be considered too disruptive to the household.

Section B summarizes indicators of the socioeconomic characteristic of migrants. Prior work suggests that climate shocks are associated with higher rates of outmigration, especially among those who are poor and engaged in agriculture (Carrico & Donato, 2019; Sedova & Kalkuhl, 2020). These data partially reflect that pattern. In disaster-affected years, the pool of outbound migrants was disproportionately more likely to come from households engaged in agriculture in the year prior—as indicated by the head's livelihood—relative to unaffected years (16% vs 11%). However, we also see evidence that migrants more often came from households engaged in business or professional work in the year prior to affected (34%) vs unaffected (27%) years. These analyses suggest no systematic differences in literacy, educational attainment, and working in the unskilled sector.

Section C includes several health indicators in an attempt to explore the extent to which environmental shocks may differentially select for migrants as a function of their health at the time of the event. Prior work suggests that in many settings migrants are healthier on average than non-migrants living in origin communities (Cunningham et al., 2008; Kennedy et al., 2015; Lu, 2008; Lu & Qin, 2014; Riosmena et al., 2013). This is true even controlling for potential confounders such as age, gender, and socioeconomic status and has been observed despite the fact that many migrants have limited access to health services in the destination. The question of whether climate shocks could alter this "healthy migrant effect" has been raised (Hunter & Simon, 2017). If environmental stress compromises livelihoods in origins, less healthy individuals may migrate alongside their healthy peers out of necessity. Alternatively, environmental stress could introduce further barriers to executing a migrant trip that exacerbates the disadvantages faced by less healthy individuals, thereby intensifying the healthy migrant effect. Mixed evidence for these competing hypotheses has been found using data from the Mexican Migration Project about international trips to the United States (Hunter & Simon, 2017), including some evidence that rainfall shocks may weaken the healthy migrant effect in moderately dry regions of Mexico.

Borrowing from the approach of Hunter and Simon (2017), we use a measure of height for heads of household and spouses, which was objectively assessed at the time of the survey. It is important to note that height is an imperfect indicator. As children develop, genetics exert a strong influence in determining height into adulthood (Silventoinen et al., 2008). However, several studies find that adult height correlates with early childhood nutrition and health, which also relates to a range of health outcomes across the life span (Alacevich & Tarozzi, 2017; Blackwell et al., 2001; Gluckman et al., 2008; Perkins et al., 2016). These relationships are small and this indicator should be interpreted carefully.

Because the BEMS only collected height measurements from heads and spouses, I constrain this analysis to migrant trips taken by these individuals. In addition, I add fixed effects representing the respondent's gender and year of birth (in 10-year increments) to capture variation in height as a person ages. I also include two measures of self-reported health. Heads and spouses were asked to describe their health at two earlier points in time: at the time of their marriage and before their first migrant trip. These measures are also imperfect but offer another window into a respondent's health status prior to their migrant trip. For all three questions, response options were poor, fair, and healthy. Because relatively few respondents reported poor health on several of these measures, I combined the response options "unhealthy" and "fairly healthy" and assigned those a value of "0." Respondents who reported being "healthy" were assigned a value of "1." The analysis involving health at the time of their marriage was constrained to trips taken by heads and spouses after their first marriage and added controls for their age at marriage and gender. The analysis for self-rated health before a first trip controlled for age at first trip and gender.

Together these analyses provide little evidence that disaster-affected years influenced the healthy migrant effect. Height and self-rated health before the first trip were roughly equivalent between affected and unaffected years. Those who migrated during disaster-affected years reported better health at marriage, with 84% of migrants during disaster-affected years reporting good health at the time of their marriage, compared to 78% in unaffected years. However, the difference was only marginally significant.

Finally, Section D considers characteristics of the migrant trips including the destination, duration, whether the migrant had ties to other migrants at the time of the trip, whether it was a first or subsequent trip, and the migrant's livelihood activity in the destination. Overall, there are surprisingly few differences across these variables. Contrary to some other findings, we see no evidence that disaster-affected years result in more internal migrants. However, trips to the capital city of Dhaka where somewhat less common (36% of disaster-affected trips vs 41%), though not significantly different. There was also no indication that environmental shocks are generating more first-time migrants. However, we do see some patterns worthy of further exploration. Those who migrated after a disaster were somewhat more likely to engage in agriculture in the destination (2% vs 1%, $p = 0.165$). Those migrants also stayed away for longer: 52% of disaster-affected migrants stayed in the destination for at least 5 years compared to 48% in unaffected years ($p = 0.157$). Although not statistically significant, the direction of this trend is counter to several findings and again suggests the need for more research using data that captures detailed information about trips.

Integrating Climate Migration Studies with Family Studies

In conclusion, there are several themes within the climate migration literature that have special relevance to the field of family studies and present opportunities for greater integration between these two fields of scholarship. First, the predominant theoretical orientation within climate migration studies considers the family to be the most relevant unit of analysis for understanding migration decisions. However, relatively few studies consider intra-household dynamics in models predicting environmentally related migration, and even fewer examine how the impacts of climate migration are felt and distributed within families. Nevertheless, a shared recognition of the primacy of families across the two fields can and should be leveraged in future work. Within the climate migration literature, this orientation often results in approaches that sample at the household level. Surveys commonly capture data about all or multiple household members to understand the migration processes within the context of household dynamics and capital resources. At a most basic level, the data generated have wide relevance for scholarship in family studies, offering a window into family dynamics in the face of mounting climate pressure and the social change that results. Questions regarding how climate change might shift social roles, norms, and responsibilities within households have been raised but only minimally addressed. Likewise, persistent heterogeneity in the relationship between climate variability and migration patterns suggests the need for more disaggregated analysis that would benefit from the perspectives and approaches offered through family studies.

Second, the simple descriptive analysis presented here, along with recent findings in the climate and migration literature, reveals a complex story about the implications for families. The data described suggest that disasters select for migrants who are on average older, shoulder more household responsibilities, and are engaged in high-risk agricultural livelihoods. Mounting evidence suggests that disruptions to resource-dependent livelihoods drive climate migration throughout much of the Global South. Mobility should, therefore, be understood as an important adaptation strategy but one that may lead to significant disruptions for families with uncertain payoffs. The suggestion that disaster-affected migrants may stay in destinations for longer and are less likely to move to the economic center of Dhaka also raises unanswered question about the outcomes of these trips that should be considered in future analyses.

Finally, intersections between scholarship on migration and adaptation have also primed important questions about when and why migration is pursued over other forms of adaptation. Prior work reveals that migration is often used as a last resort to cope with environmental impacts (Jülich, 2011; Sakdapolrak et al., 2014). Other forms of adaptation, when available, may shift the costs, benefits, and risks of sending a migrant. Alternatively, migration may also offset the need to pursue adaptation strategies that result in negative long-term social and economic outcomes, such as removing children from school (Randell & Gray, 2019), marrying a daughter (Carrico et al., 2020), or selling assets (Mortimore, 2010). In short, it is critical to understand migration decisions within the context of the diverse risks and resources that families experience.

Acknowledgments This research was funded by the National Science Foundation (CNH-1716909 and BCS-2149191) and the Office of Naval Research (#N00014-11-1-0683). This manuscript was also supported by the University of Colorado Population Center (Project 2P2CHD066613-11), funded by the Eunice Kennedy Shriver National Institute of Child Health and Human Development. I wish to thank Katharine Donato for comments on this analysis.

References

Alacevich, C., & Tarozzi, A. (2017). Child height and intergenerational transmission of health: Evidence from ethnic Indians in England. *Economics & Human Biology, 25*, 65–84. https://doi.org/10.1016/j.ehb.2016.10.004

Angelsen, A., Aguilar-Støen, M., Ainembabazi, J., Castellanos, E., & Taylor, M. (2020). Migration, remittances, and forest cover change in rural Guatemala and Chiapas, Mexico. *Land, 9*(3), 88. https://doi.org/10.3390/land9030088

Auriat, N. (1991). Who forgets? An analysis of memory effects in a retrospective survey on migration history. *European Journal of Population, 7*(4), 311–342. https://doi.org/10.1007/BF01796872

Baird, R., Migiro, K., Nutt, D., Kwatra, A., Wilson, S. J., Melby, J., Pendleton, A., Rodgers, M., & Davison, J. (2007, May 14). Human tide: The real migration crisis. *Christian Aid Report*. https://www.christianaid.org.uk/sites/default/files/2017-08/human-tide-the-real-migration-crisis-may-2007.pdf

Barnett, J., & Adger, W. N. (2007). Climate change, human security and violent conflict. *Political Geography, 26*(6), 639–655. https://doi.org/10.1016/j.polgeo.2007.03.003

Basak, J. K., Ali, M. A., & Islam, N. (2010). Assessment of the effect of climate change on boro rice production in Bangladesh using DSSAT model. *Journal of Civil Engineering (IEB), 38*(2), 95–108.

Bebbington, A. (1999). Capitals and capabilities: A framework for analyzing peasant viability, rural livelihoods and poverty. *World Development, 27*(12), 2021–2044. https://doi.org/10.1016/S0305-750X(99)00104-7

Berchin, I. I., Valduga, I. B., Garcia, J., & de Andrade Guerra, J. B. S. O. (2017). Climate change and forced migrations: An effort towards recognizing climate refugees. *Geoforum, 84*, 147–150. https://doi.org/10.1016/j.geoforum.2017.06.022

Biden, J. R., Jr. (2021, Febraury 24). *Rebuilding and enhancing programs to resettle refugees and planning for the impact of climate change on migration* (executive order no. 14013). The White House, Briefing Room. https://www.whitehouse.gov/briefing-room/presidential-actions/2021/02/04/executive-order-on-rebuilding-and-enhancing-programs-to-resettle-refugees-and-planning-for-the-impact-of-climate-change-on-migration/

Black, R., Adger, W. N., Arnell, N. W., Dercon, S., Geddes, A., & Thomas, D. (2011). The effect of environmental change on human migration. *Global Environmental Change, 21*, S3–S11. https://doi.org/10.1016/j.gloenvcha.2011.10.001

Blackwell, D. L., Hayward, M. D., & Crimmins, E. M. (2001). Does childhood health affect chronic morbidity in later life? *Social Science & Medicine, 52*(8), 1269–1284. https://doi.org/10.1016/S0277-9536(00)00230-6

Bohra-Mishra, P. (2013). Labour migration and investments by remaining households in rural Nepal. *Journal of Population Research, 30*(2), 171–192. https://doi.org/10.1007/s12546-012-9097-2

Bohra-Mishra, P., Oppenheimer, M., & Hsiang, S. M. (2014). Nonlinear permanent migration response to climatic variations but minimal response to disasters. *Proceedings of the National Academy of Sciences, 111*(27), 9780–9785. https://doi.org/10.1073/pnas.1317166111

Call, M., & Gray, C. (2020). Climate anomalies, land degradation, and rural out-migration in Uganda. *Population and Environment, 41*(4), 507–528. https://doi.org/10.1007/s11111-020-00349-3

Call, M., Gray, C., Yunus, M., & Emch, M. (2017). Disruption, not displacement: Environmental variability and temporary migration in Bangladesh. *Global Environmental Change, 46*, 157–165. https://doi.org/10.1016/j.gloenvcha.2017.08.008

Carrico, A. R., & Donato, K. (2019). Extreme weather and migration: Evidence from Bangladesh. *Population and Environment, 41*(1), 1–31. https://doi.org/10.1007/s11111-019-00322-9

Carrico, A. R., Donato, K. M., Best, K. B., & Gilligan, J. (2020). Extreme weather and marriage among girls and women in Bangladesh. *Global Environmental Change, 65*, 102160. https://doi.org/10.1016/j.gloenvcha.2020.102160

Center for Research on The Epidemiology of Disasters (CRED). (2021). *EM-DAT: The international disaster database* [Database]. Accessed Aug. 23, 2021. EM-DAT | The international disasters database (emdat.be)

Chazalnoel, M. T., & Ionesco, D. (2016, June 6). *Defining climate migrants – beyond semantics*. ImmigrationProf Blog. https://lawprofessors.typepad.com/immigration/2016/06/defining-climate-migrants-beyond-semantics.html

Chen, J., & Mueller, V. (2018). Coastal climate change, soil salinity and human migration in Bangladesh. *Nature Climate Change, 8*(11), 981–985. https://doi.org/10.1038/s41558-018-0313-8

Chen, J. J., Mueller, V., Jia, Y., & Tseng, S. K.-H. (2017). Validating migration responses to flooding using satellite and vital registration data. *The American Economic Review, 107*(5), 441–445.

Chiswick, B. R., Lee, Y. L., & Miller, P. W. (2008). Immigrant selection systems and immigrant health. *Contemporary Economic Policy, 26*(4), 555–578. https://doi.org/10.1111/j.1465-7287.2008.00099.x

Cunningham, S. A., Ruben, J. D., & Venkat Narayan, K. M. (2008). Health of foreign-born people in the United States: A review. *Health & Place, 14*(4), 623–635. https://doi.org/10.1016/j.healthplace.2007.12.002

Dell, M., Jones, B., & Olken, B. (2008). *Climate change and economic growth: Evidence from the last half century* (Report no. w14132). National Bureau of Economic Research. https://doi.org/10.3386/w14132

Dillon, A., Mueller, V., & Salau, S. (2011). Migratory responses to agricultural risk in northern Nigeria. *American Journal of Agricultural Economics, 93*(4), 1048–1061. https://doi.org/10.1093/ajae/aar033

Donato, K. M., Carrico, A. R., Sisk, B., & Piya, B. (2016). Different but the same: How legal status affects international migration from Bangladesh. *The Annals of the American Academy of Political and Social Science, 666*(1), 203–218. https://doi.org/10.1177/0002716216650843

Dun, O. V., & Gemenne, F. (2008). Defining "environmental migration." *Forced Migration Reviews, 31*, 10–11. https://www.fmreview.org/climatechange/dun-gemenne

Eklund, L., Romankiewicz, C., & Brandt, M. (2016). *Data and methods in the environment-migration nexus: A scale perspective* (2nd ed.). Gesellschaft für Erdkunde zu Berlin. https://doi.org/10.12854/erde-147-10

Fiorino, D. J. (2018). *Can democracy handle climate change?* Wiley.

Freedman, D., Thornton, A., Camburn, D., Alwin, D., & Young-DeMarco, L. (1988). The life history calendar: A technique for collecting retrospective data. *Sociological Methodology, 18*, 37–68. https://doi.org/10.2307/271044

Fussell, E., Hunter, L. M., & Gray, C. L. (2014). Measuring the environmental dimensions of human migration: The demographer's toolkit. *Global Environmental Change, 28*, 182–191. https://doi.org/10.1016/j.gloenvcha.2014.07.001

Gemenne, F. (2011). Why the numbers don't add up: A review of estimates and predictions of people displaced by environmental changes. *Global Environmental Change, 21S*, S41–S49. https://doi.org/10.1016/j.gloenvcha.2011.09.005

Gluckman, P. D., Hanson, M. A., Cooper, C., & Thornburg, K. L. (2008). Effect of in utero and early-life conditions on adult health and disease. *New England Journal of Medicine, 359*(1), 61–73. https://doi.org/10.1056/NEJMra0708473

Gray, C. L., & Mueller, V. (2012a). Natural disasters and population mobility in Bangladesh. *PNAS, 109*(16), 6000–6005. https://doi.org/10.1073/pnas.1115944109

Gray, C., & Mueller, V. (2012b). Drought and population mobility in rural Ethiopia. *World Development, 40*(1), 134–145. https://doi.org/10.1016/j.worlddev.2011.05.023

Grothmann, T., & Patt, A. (2005). Adaptive capacity and human cognition: The process of individual adaptation to climate change. *Global Environmental Change, 15*(3), 199–213. https://doi.org/10.1016/j.gloenvcha.2005.01.002

Henry, S., Schoumaker, B., & Beauchemin, C. (2004). The impact of rainfall on the first out-migration: A multi-level event-history analysis in Burkina Faso. *Population and Environment, 25*(5), 423–460. https://doi.org/10.1023/B:POEN.0000036928.17696.e8

Hoffmann, R., Dimitrova, A., Muttarak, R., Crespo Cuaresma, J., & Peisker, J. (2020). A meta-analysis of country-level studies on environmental change and migration. *Nature Climate Change, 10*(10), 904–912. https://doi.org/10.1038/s41558-020-0898-6

Hoffmann, R., Šedová, B., & Vinke, K. (2021). Improving the evidence base: A methodological review of the quantitative climate migration literature. *Global Environmental Change, 71*, 102367. https://doi.org/10.1016/j.gloenvcha.2021.102367

Hugo, G. (1996). Environmental concerns and international migration. *The International Migration Review, 30*(1), 105–131. https://doi.org/10.2307/2547462

Hunter, L. M. (2018). Where do we go from here? Reflections on the future of environmental migration and displacement research. In R. McLeman & F. Gemenne (Eds.), *Routledge handbook of environmental displacement and migration* (pp. 430–435). Routledge.

Hunter, L. M., & Simon, D. H. (2017). Might climate change the "healthy migrant" effect? *Global Environmental Change, 47*, 133–142. https://doi.org/10.1016/j.gloenvcha.2017.10.003

Hunter, L. M., Murray, S., & Riosmena, F. (2013). Rainfall patterns and U.S. migration from rural Mexico. *International Migration Review, 47*(4), 874–909. https://doi.org/10.1111/imre.12051

International Organization for Migration (IOM). (2019). *International migration law: Glossary on migration* (No. 34). https://publications.iom.int/system/files/pdf/iml_34_glossary.pdf

International Organization for Migration (IOM) (2021). *Institutional strategy on migration, environment, and climate change 2021–2030*. IOM-Institutional-Strategy-MECCC_0.pdf

IPCC (2018). *Summary for policymakers. Special report: Global warming of 1.5°C*. https://www.ipcc.ch/sr15/chapter/spm/

Jacobson, J. L. (1988). Environmental refugees: A yardstick of habitability. *Bulletin of Science, Technology & Society, 8*(3), 257–258. https://doi.org/10.1177/027046768800800304

Jülich, S. (2011). Drought triggered temporary migration in an east Indian village. *International Migration, 49*, e189–e199. https://doi.org/10.1111/j.1468-2435.2010.00655.x

Kaczan, D. J., & Orgill-Meyer, J. (2020). The impact of climate change on migration: A synthesis of recent empirical insights. *Climatic Change, 158*(3), 281–300. https://doi.org/10.1007/s10584-019-02560-0

Kennedy, S., Kidd, M. P., McDonald, J. T., & Biddle, N. (2015). The healthy immigrant effect: Patterns and evidence from four countries. *Journal of International Migration and Integration, 16*(2), 317–332. https://doi.org/10.1007/s12134-014-0340-x

Lilleør, H. B., & Van den Broeck, K. (2011). Economic drivers of migration and climate change in LDCs. *Global Environmental Change, 21*(Suppl. 1), S70–S81. https://doi.org/10.1016/j.gloenvcha.2011.09.002

Lindstrom, D. P., & López Ramírez, A. (2010). Pioneers and followers: Migrant selectivity and the development of U.S. migration streams in Latin America. *The Annals of the American Academy of Political and Social Science, 630*(1), 53–77. https://doi.org/10.1177/0002716210368103

Lu, Y. (2008). Test of the 'healthy migrant hypothesis': A longitudinal analysis of health selectivity of internal migration in Indonesia. *Social Science & Medicine, 67*(8), 1331–1339. https://doi.org/10.1016/j.socscimed.2008.06.017

Lu, Y., & Qin, L. (2014). Healthy migrant and salmon bias hypotheses: A study of health and internal migration in China. *Social Science & Medicine, 102*, 41–48. https://doi.org/10.1016/j.socscimed.2013.11.040

Mallick, B., & Etzold, B. (2015). *Environment, migration and adaptation: Evidence and politics of climate change in Bangladesh*. A H Development Publishing House.

Mallick, B., & Vogt, J. (2012). Cyclone, coastal society and migration: Empirical evidence from Bangladesh. *International Development Planning Review, 34*(3), 217–240. https://doi.org/10.3828/idpr.2012.16

Massey, D. S., Arango, J., Hugo, G., Kouaouci, A., & Pellegrino, A. (1999). *Worlds in motion: Understanding international migration at the end of the millennium*. Clarendon Press.

McAdam, J. (2010). *Climate change and displacement: Multidisciplinary perspectives*. Bloomsbury Publishing.

Moniruzzaman, M. (2020). The impact of remittances on household food security: Evidence from a survey in Bangladesh. *Migration and Development*, 1–20. https://doi.org/10.1080/21632324.2020.1787097

Mortimore, M. (2010). Adapting to drought in the Sahel: Lessons for climate change. *Wiley Interdisciplinary Reviews: Climate Change, 1*(1), 134–143. https://doi.org/10.1002/wcc.25

Mueller, V., Gray, C., & Kosec, K. (2014). Heat stress increases long-term human migration in rural Pakistan. *Nature Climate Change, 4*(3), 182–185. https://doi.org/10.1038/nclimate2103

Myers, N. (2002). Environmental refugees: A growing phenomenon of the 21st century. *Philosophical Transactions of the Royal Society B: Biological Sciences, 357*(1420), 609–613. https://doi.org/10.1098/rstb.2001.0953

Nawrotzki, R. J., & Bakhtsiyarava, M. (2017). International climate migration: Evidence for the climate inhibitor mechanism and the agricultural pathway. *Population, Space and Place, 23*(4), e2033. https://doi.org/10.1002/psp.2033

Nawrotzki, R. J., & DeWaard, J. (2016). Climate shocks and the timing of migration from Mexico. *Population and Environment, 38*(1), 72–100. https://doi.org/10.1007/s11111-016-0255-x

Nawrotzki, R. J., & DeWaard, J. (2018). Putting trapped populations into place: Climate change and inter-district migration flows in Zambia. *Regional Environmental Change, 18*(2), 533–546.

Nawrotzki, R. J., Riosmena, F., Hunter, L. M., & Runfola, D. M. (2015). Amplification or suppression: Social networks and the climate change—Migration association in rural Mexico. *Global Environmental Change, 35*, 463–474. https://doi.org/10.1016/j.gloenvcha.2015.09.002

Ngoma, A. L., & Ismail, N. W. (2013). Do migrant remittances promote human capital formation? Evidence from 89 developing countries. *Migration and Development, 2*(1), 106–116. https://doi.org/10.1080/21632324.2013.786883

Palloni, A., & Morenoff, J. D. (2006). Interpreting the paradoxical in the Hispanic paradox: Demographic and epidemiologic approaches. *Annals of the New York Academy of Sciences, 954*(1), 140–174. https://doi.org/10.1111/j.1749-6632.2001.tb02751.x

Panda, A. (2017). Vulnerability to climate variability and drought among small and marginal farmers: A case study in Odisha, India. *Climate and Development, 9*(7), 605–617. https://doi.org/10.1080/17565529.2016.1184606

Perkins, J. M., Subramanian, S. V., Davey Smith, G., & Özaltin, E. (2016). Adult height, nutrition, and population health. *Nutrition Reviews, 74*(3), 149–165.

Piguet, E. (2010). Linking climate change, environmental degradation, and migration: A methodological overview. *WIREs Climate Change, 1*(4), 517–524. https://doi.org/10.1002/wcc.54

Rahman, M. H., Lund, T., & Bryceson, I. (2011). Salinity impacts on agro-biodiversity in three coastal, rural villages of Bangladesh. *Ocean & Coastal Management, 54*(6), 455–468. https://doi.org/10.1016/j.ocecoaman.2011.03.003

Randell, H., & Gray, C. (2019). Climate change and educational attainment in the global tropics. *Proceedings of the National Academy of Sciences, 116*(18), 8840–8845. https://doi.org/10.1073/pnas.1817480116

Renaud, F. G., Dun, O., Warner, K., & Bogardi, J. (2011). A decision framework for environmentally induced migration. *International Migration, 49*(s1), e5–e29. https://doi.org/10.1111/j.1468-2435.2010.00678.x

Riosmena, F., Wong, R., & Palloni, A. (2013). Migration selection, protection, and acculturation in health: A binational perspective on older adults. *Demography, 50*(3), 1039–1064. https://doi.org/10.1007/s13524-012-0178-9

Robalino, J., Jimenez, J., & Chacón, A. (2015). The effect of hydro-meteorological emergencies on internal migration. *World Development, 67*, 438–448. https://doi.org/10.1016/j.worlddev.2014.10.031

Saha, S. K. (2017). Cyclone Aila, livelihood stress, and migration: Empirical evidence from coastal Bangladesh. *Disasters, 41*(3), 505–526. https://doi.org/10.1111/disa.12214

Sakdapolrak, P., Promburom, P., & Reif, A. (2014). Why successful *in situ* adaptation with environmental stress does not prevent people from migrating? Empirical evidence from northern Thailand. *Climate and Development, 6*(1), 38–45. https://doi.org/10.1080/17565529.2013.826129

Sedova, B., & Kalkuhl, M. (2020). Who are the climate migrants and where do they go? Evidence from rural India. *World Development, 129*, 104848. https://doi.org/10.1016/j.worlddev.2019.104848

Sikder, M. J. U., & Higgins, V. (2017). Remittances and social resilience of migrant households in rural Bangladesh. *Migration and Development, 6*(2), 253–275. https://doi.org/10.1080/21632324.2016.1142752

Silventoinen, K., Pietiläinen, K. H., Tynelius, P., Sørensen, T. I. A., Kaprio, J., & Rasmussen, F. (2008). Genetic regulation of growth from birth to 18 years of age: The Swedish young male twins study. *American Journal of Human Biology, 20*(3), 292–298. https://doi.org/10.1002/ajhb.20717

Smith, J. P., & Thomas, D. (2003). Remembrances of things past: Test–retest reliability of retrospective migration histories. *Journal of the Royal Statistical Society: Series A (Statistics in Society), 166*(1), 23–49. https://doi.org/10.1111/1467-985X.00257

Stark, O., & Bloom, D. E. (1985). The new economics of labor migration. *The American Economic Review, 75*(2), 173–178. https://www.jstor.org/stable/1805591

Streatfield, P. K., & Karar, Z. A. (2008). Population challenges for Bangladesh in the coming decades. *Journal of Health, Population and Nutrition, 26*(3), 261–272. https://doi.org/10.3329/jhpn.v26i3.1894

Tacoli, C. (2009). Crisis or adaptation? Migration and climate change in a context of high mobility. *Environment and Urbanization, 21*(2), 513–525. https://doi.org/10.1177/0956247809342182

Thiede, B., Gray, C., & Mueller, V. (2016). Climate variability and inter-provincial migration in South America, 1970–2011. *Global Environmental Change, 41*, 228–240. https://doi.org/10.1016/j.gloenvcha.2016.10.005

Truelove, H. B., Carrico, A. R., & Thabrew, L. (2015). A socio-psychological model for analyzing climate change adaptation: A case study of Sri Lankan paddy farmers. *Global Environmental Change, 31*, 85–97. https://doi.org/10.1016/j.gloenvcha.2014.12.010

Wilmsen, B., & Webber, M. (2015). What can we learn from the practice of development-forced displacement and resettlement for organised resettlements in response to climate change? *Geoforum, 58*, 76–85. https://doi.org/10.1016/j.geoforum.2014.10.016

World Bank. (2020). *Personal remittances, received (% of GDP)*. DataBank. https://data.worldbank.org/indicator/BX.TRF.PWKR.DT.GD.ZS?year_high_desc=true

Zickgraf, C. (2021). Climate change, slow onset events and human mobility: Reviewing the evidence. *Current Opinion in Environmental Sustainability, 50*, 21–30. https://doi.org/10.1016/j.cosust.2020.11.007

Part III
The Built Environment

Chapter 5
Built Environment, Family Processes, and Child and Adolescent Health and Well-Being

Kim T. Ferguson and Gary W. Evans

Although psychology and related disciplines long neglected the impacts of the physical environment on human development, in the last 50 years, a robust body of multidisciplinary work has documented the impacts of the physical environment, including the built environment, on adult health and well-being (see, e.g., Browning et al., 2011; Clayton, 2012; Gifford, 2014; Steg & De Groot, 2019; Wilkie et al., 2018). The built environment refers to places and spaces created and/or modified by people, including built structures, parks, and transportation systems. In addition, the literature on the impacts of the physical environment on child health and well-being, particularly in the Global North, has grown significantly in the last two decades (see, e.g., Christian et al., 2015; Evans, 2006, 2021; Ferguson et al., 2013; Whipple & Evans, 2022; Wigle, 2003). There is, nevertheless, less work to date documenting these impacts on those growing up in the Global South and in underrepresented communities (e.g., low-income children, BIPOC [black, indigenous, and people of color] children, rural children, displaced and unhoused children) in the Global North (Ferguson et al., 2013). Furthermore, much of the work to date has documented direct impacts of the physical environment on individual children, adolescents, and adults at specific points in developmental and historical time. Longitudinal, multilevel work remains limited, as does work that directly pinpoints key mechanisms of action underlying how key aspects of the physical environment individually and jointly impact specific components of child well-being (Doan & Evans, 2020).

K. T. Ferguson (✉)
Sarah Lawrence College, Bronxville, NY, USA
e-mail: kferguson@sarahlawrence.edu

G. W. Evans
College of Human Ecology, Cornell University, Ithaca, NY, USA
e-mail: gwe1@cornell.edu

S. E. Ortiz et al. (eds.), *Environmental Impacts on Families*, National Symposium on Family Issues 12, https://doi.org/10.1007/978-3-031-22649-6_5

In this chapter, we focus on the impacts of the physical environment (toxins and pollutants, food and water insecurity, lighting and temperature, noise, crowding, chaos, and housing quality) on family processes and on interactions of children and adolescents within home environments and extensions of home (nearby play spaces, afterschool settings, and youth centers). We also briefly discuss impacts on two extensions of home, child and sibling school and childcare environments and sibling and caregiver work environments. We focus specifically on several family processes that are known to be impacted by key aspects of the physical environment, namely, parenting practices, attachment, caregiver-child relationships and interactions, sibling relationships, family routines and rituals, family conflict, interpersonal aggression, child maltreatment, and social support (including social capital and social networks). We consider both the spaces that children, adolescents, and their families inhabit and the materials that make up those spaces and are embedded within them, such as structural quality, ambient conditions, food, water, books, and play materials outside of the home. We also discuss both the opportunities and constraints (affordances; Gibson, 1979) of various spaces and materials. Following this review, we discuss needed actions for change based on the extant literature, gaps in the work to date, and potential avenues for future research. We pay particular attention to racial and socioeconomic inequalities in both exposure to adverse environmental conditions and access to positive physical resources.

Family, Household, and Home

To understand the impacts of the physical environment on child-family interactions and processes, we define *family*, *household,* and *home* in the context of the present discussion. In this chapter, we primarily consider interactions between family members currently living within the same household, with a specific focus on interactions between caregivers and their dependents and between siblings living within the same household. Of course, these definitions are imperfect given the heterogeneity of family systems (see, e.g., Fiese et al., 2019), particularly when varying conceptualizations of family across the Global South are taken into consideration (see Bernal et al., 2019; Ferguson & Evans, 2019; Ferguson & Lee, 2013; Nsamenang, 1992; Serpell, 2011). When possible, we also discuss relevant research on family interactions for children living in multiple households on a regular basis, households made up of multiple family units and/or multiple separate dwelling spaces, and multigenerational households within the same home or compound, including shared home care facilities for orphaned and vulnerable children and adolescents in the Global South. Throughout, we extend the concept of home to the nearby spaces that children engage in such as streets, parks, fields, and other outdoor spaces. We use the convention child to refer to individuals from birth through age 18, specifying child or adolescent when salient family processes vary with maturation.

A Bioecocultural Framework for Mapping Family Processes

To map out the various impacts, both direct and indirect, of the physical environment on family interactions and processes, we draw on an extension of Bronfenbrenner's (Bronfenbrenner, 1977) bioecological model (see also Bronfenbrenner & Ceci, 1994; Bronfenbrenner & Crouter, 1983; Bronfenbrenner & Evans, 2000; Bronfenbrenner & Morris, 2006), a bioecocultural model that provides a conceptual framework for studying human development and family systems in context (Ferguson & Evans, 2019; see Fig. 5.1).

As detailed in Fig. 5.1, the developing child or adolescent is embedded within family, neighborhood, school, and broader sociocultural settings, with family interactions and processes, including family routines and rituals, operating as key "engines of development" (Bronfenbrenner & Morris, 2006; Ferguson & Evans, 2019; Fiese, 2006; Weisner, 2002, 2010). These *proximal processes*, occurring regularly and sustaining reciprocal interactions between the developing *person* and the persons, objects, and symbols within their immediate physical and social *context*, drive the development of the child over developmental and historical *time* (see Fig. 5.1 and Bronfenbrenner & Evans, 2000; Bronfenbrenner & Morris, 2006). Importantly, proximal processes are most effective when they become increasingly

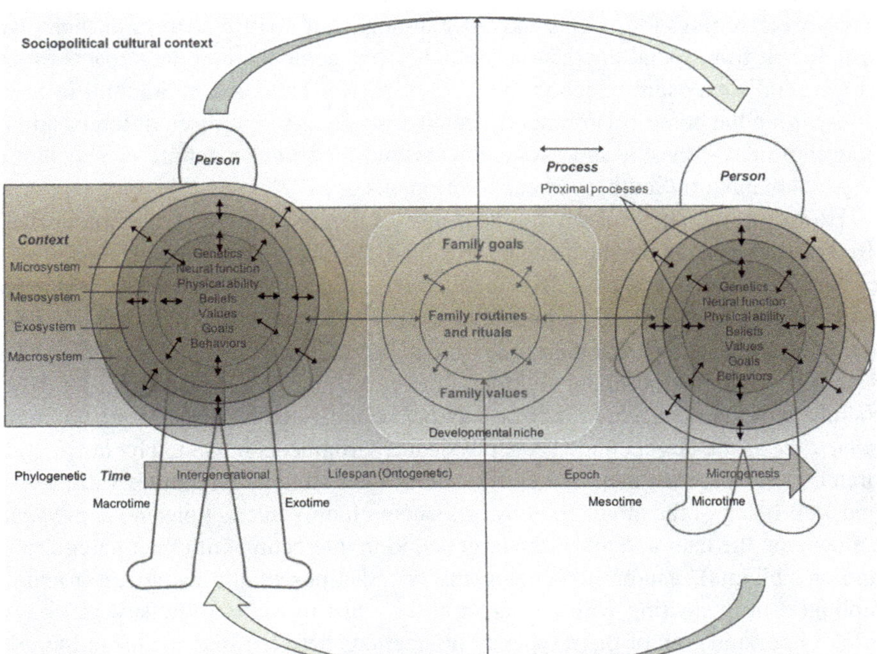

Fig. 5.1 Bronfenbrenner's bioecological model re-envisioned: Extending Process, Person, Context, and Time (PPCT) to create a bioecocultural model. (Reprinted with permission from Ferguson and Evans (2019))

more complex with maturation. Their power also varies as a function of the characteristics of the developing *person,* including demand characteristics such as gender that influence family interactions, resources, and behavioral dispositions (Ferguson & Evans, 2019). Over developmental and historical *time,* such differences in family interactions (*processes*) themselves influence the characteristics of the developing *person,* including their strengths, abilities, and beliefs.

As can be seen in Fig. 5.1, the influence of the physical and social environments in which the developing child and their family are embedded are conceptualized as impacting the developing person both directly (within the *microsystem* or the immediate environments directly experienced by the developing person, including the family, home, household, peer group, school, and neighborhood) and indirectly (within the *mesosystem, exosystem*, and *macrosystem*; see also Bronfenbrenner & Morris, 2006). The mesosystem consists of intersections between microsystems in which the developing person is embedded, such as relationships between the home and neighborhood settings, and reflects the capacity of proximal processes within one microsystem to impact those in another (Bronfenbrenner & Morris, 2006). For example, children exposed to noise in school may be more likely to withdraw from social interactions later, when they return home, especially if their homes are also noisy and/or otherwise chaotic. The *exosystem* includes intersections between settings that may not contain the developing person, such as parental work settings and sibling play settings that are not entered by the developing child on a regular basis. Noisy work settings for parents and older siblings may contribute towards their own withdrawal from social interactions at home, thus compounding the impacts of the target child's exposure to school noise on caregiver-child and child-child interactions within the home environment. The macrosystem is comprised of intersections between micro-, meso-, and exosystems within a particular culture or subculture (e.g., Global North/South, socioeconomic status or SES).

How does the physical environment impact family interactions and processes? To better understand the mechanisms involved, consider that the target child spends part of their time in contexts also inhabited by caregivers and siblings, part of their time in contexts shared only with siblings, part of their time in contexts shared only with caregivers, and part of their time in contexts seldom or never inhabited by other members of their household. Furthermore, the contexts also inhabited by other members of their household may or may not be inhabited by those members *at the same time* as the target child. Thus, physical environments inhabited by target children and adolescents, their caregivers, and/or their siblings might affect the health and well-being of the target child by influencing family interactions *in the moment,* influencing the later actions of the target child in interacting with their caregiver(s) and/or sibling(s), and/or influencing the later actions of the caregiver(s) and/or sibling(s) in interacting with the target child – just to name a few key pathways. And, importantly, all of these types of interactions between individuals and family groups and their shared and unshared physical environments can take place idiosyncratically or regularly and over shorter or longer periods of time. Thus, all four dimensions of Bronfenbrenner's bioecological model, namely, *person, process, context*, and *time*, are important to consider in better understanding the impacts of the physical environment on a target child or adolescent. Figure 5.2 illustrates home, neighborhood, school, childcare, play, afterschool, youth center, and work settings with potential movements of children of different ages and their caregiver(s) within and between these settings. It should be noted here that this is only one example of a potential household, home, and family structure. These structures, and thus the relevant spaces occupied by relevant close family members, might vary considerably in, for example, a multiple household family structure in the Global North or South.

Evans' (2021, Fig. 1) preliminary taxonomy of physical environment characteristics and child development highlights the importance of considering the scale of a setting or the proximity of the environmental characteristic to the child (proximal:

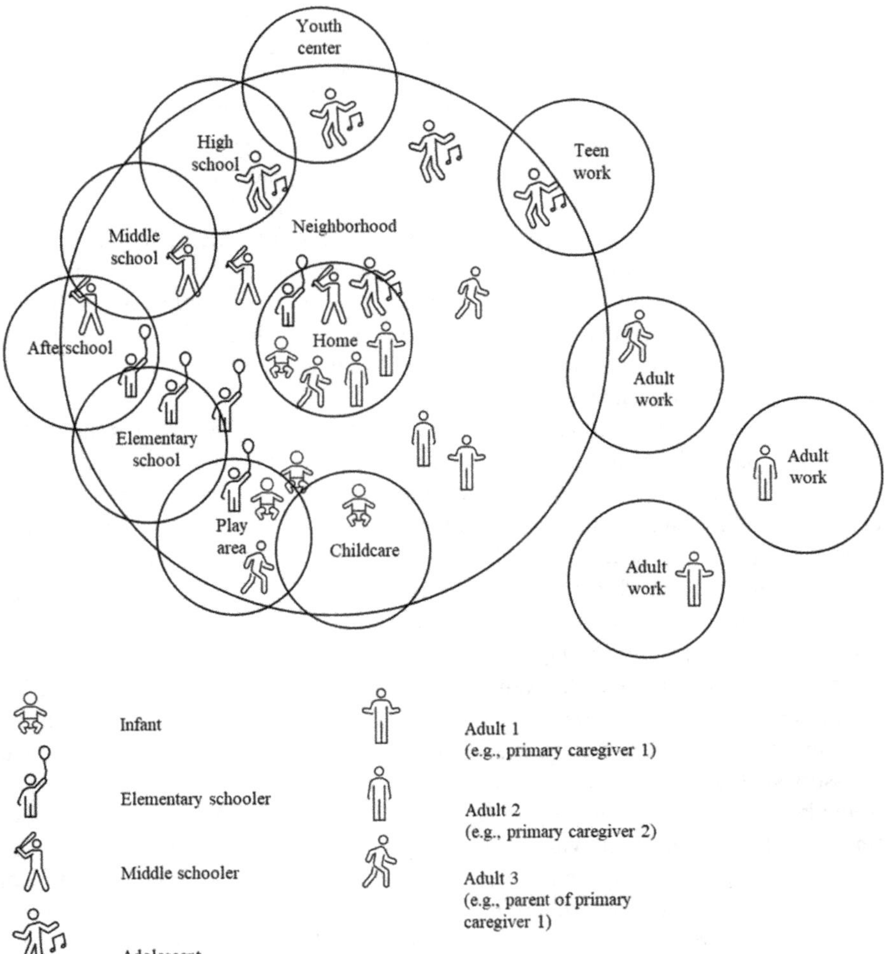

Fig. 5.2 Home, extensions of home, and beyond home: Key physical settings regularly inhabited by children and adults within a sample household

home, school, childcare; medial: neighborhood, playground, teen center; distal: national, Global); direct versus indirect effects (experienced by the child themselves or by others such as caregivers, siblings, and teachers operating within the child's microsystem); and the temporal dynamics of the child-environment interaction (including duration, developmental timing, and the regularity of interaction).

In this chapter, we focus primarily on proximal and some medial characteristics and on both direct and indirect effects. When possible, we discuss what is known of temporal dynamics. As we do so, we highlight the importance of considering the affordances (Gibson, 1979) of different physical environments, including both opportunities and constraints created by the dynamic interactions between the child, their caregivers and siblings, and the environments they inhabit. Thus, we will discuss mechanisms of action, including the creation of negative interactions (e.g., conflict, withdrawal) and the prevention of positive interactions, primarily through interruption, disruption, and reorganization of family interactions and processes as a result of suboptimal housing, noise, crowding, and environmental chaos (Bronfenbrenner & Evans, 2000; Doan & Evans, 2020; Evans, 2021; Evans & Wachs, 2010; Ferguson & Evans, 2019). For example, safe green play spaces afford more positive child-child and caregiver-child play and language interactions (Loebach & Cox, 2020; Maxwell et al., 2008; van Dijk-Wesselius et al., 2018; Yogman et al., 2018). In contrast, living on the top floor of a high-rise building in a neighborhood with limited affordances for safe play constrains child-child and caregiver-child play and language interactions and negatively impacts parent and child mood and engagement (Evans, 2003; Wells & Moch, 2003). Noisy environments impact the target child, siblings, and adult caregivers by contributing to higher levels of annoyance and irritability, and thus likely result in fewer and less sustained positive interactions and a greater number of negative interactions within the noisy household. Unpredictable, unstable, and chaotic physical and psychosocial contexts that interrupt, disrupt, and reorganize family interactions and processes, including family routines and rituals, have particularly deleterious impacts on child development, health, and well-being (Bronfenbrenner, 1986; Bronfenbrenner & Evans, 2000; Doan & Evans, 2020; Ferguson et al., 2013; Ferguson & Evans, 2019). Recent and ongoing work by Fiese and colleagues (Fiese, 2006; Fiese et al., 2015; Fiese & Schwartz, 2008; Saltzman et al., 2019) and by Evans, Wachs, and colleagues (Bronfenbrenner & Evans, 2000; Doan & Evans, 2020; Evans & Wachs, 2010; Whipple & Evans, 2022) has identified key family processes that might be interrupted, disrupted, and/or reorganized as a result of exposure to environmental stressors such as poor housing quality, noise, crowding, and environmental chaos. Some key family processes we will discuss in this chapter include caregiving routines; family mealtimes; individual play and games; shared caregiver-child activities such as reading, child play, and games; and child-family support, both instrumental (e.g., homework) and emotional (e.g., responsiveness). Outcomes to be discussed, given the extant research documenting physical environment impacts that influence child health and well-being, include parenting practices, attachment, caregiver-child relationships and interactions, sibling relationships, family routines and rituals, family conflict, parent and sibling aggression, child maltreatment, and social support (including social capital and social networks).

The Physical Environment and Family Interactions and Processes

Home

Toxins and Pollutants

Lead: Extensive evidence documents the direct effects of prenatal and childhood lead exposure on child health and cognitive and socioemotional functioning across both the Global North and South, particularly if the exposure is early and of longer duration (Evans, 2003, 2006; Ferguson et al., 2013). Further, evidence in humans and mice suggests that timing, level, and duration of lead exposure are implicated in the duration and strength of impact (Banna et al., 2022; Sampson & Winter, 2018; Searle et al., 2014).

Much less is known of the impacts of such exposure on family processes and interactions. Nevertheless, given the known impacts of lead exposure on adult cognitive and socioemotional functioning and health, including visuospatial processing, cardiovascular disease mortality, and potentially criminal behavior (e.g., Shih et al., 2006; Taylor et al., 2016), it is plausible that prior or concurrent parent exposure to lead in the home and/or at work might impact their interactions with their children. Similarly, given the impacts of lead on child behavior, including attention regulation, self-regulation, and general behavioral problems, it is quite likely that sibling exposure in the home or at work or school would similarly impact interactions with a target child. For example, reduced self-regulation is associated with behavioral conduct disorders and aggression (Robson et al., 2020), and there is some limited evidence that early lead exposure is associated with later adolescent delinquent behavior and possibly also criminal activity (Evans, 2003; Needleman et al., 1990; Sampson & Winter, 2018). Lead's potential association with later criminal behavior results in a greater likelihood of parental absence from the home, and reduced parental inhibitory control and higher irritability resulting from lead exposure may lead to more punitive and harsh parenting. Further, it has been suggested that caregivers might contribute to the lead burden of children in home environments through reduced attention, thus resulting in higher interaction with lead-contaminated household objects, but there is little direct evidence for this to date (see Searle et al., 2014). Nevertheless, evidence from animal models demonstrates increased aggression and deficient maternal behaviors following heavy metal exposure (Laughlin, 1986).

More direct evidence of potential indirect influences of family interactions on the impacts of lead exposure on child health and development has been found (Xu et al., 2015). Controlling for other relevant variables such as family SES and children's current blood lead levels, Mexican children whose mothers had higher self-esteem scores when they were toddlers had less inattention behavior at ages 7–15 years than those whose mothers had lower self-esteem scores. Most interestingly for the purposes of this chapter, the association between mother's self-esteem when their children were toddlers and child attention behaviors at ages 7–15 years was stronger among those with lower prenatal lead exposure. The finding suggests that prenatal exposure to lead reduces the positive impact of high maternal self-esteem on children's later attention behaviors. The researchers suggest, based on prior animal studies, that lead might interfere with high maternal self-esteem's modulation of child stress, noting that higher childhood stress is associated with later inattention behaviors. This being said, of course, much further research is needed to better understand the mechanism(s) of action here.

Mercury

Another heavy metal with documented direct impacts on child and adolescent motor, cognitive, and socioemotional development and later physical and mental health is mercury, with prenatal exposure being particularly problematic for later development (Barbone et al., 2019; Ferguson et al., 2013; Ng et al., 2014; Patel et al., 2019). As is the case with lead, however, there is little research to date on the impacts of exposure to methylmercury on family interactions and processes. A recent animal model indicates that mercury-exposed captive-bred zebra finches were less likely to initiate nests and spent less time constructing them and that their nests were lighter than those of controls (Chin et al., 2017). As a result, the finches were less likely to fledge young, whether the eggs were mercury-exposed or not. This study highlights the potential indirect effects on children of parental heavy metal exposure, even when the children themselves are not impacted. Nevertheless, clearly work with mice, non-human primates, and humans is warranted.

Pesticides

Extensive research in the last two decades has documented the direct impacts of (primarily dietary) prenatal and early organophosphate (OP) pesticide exposure on young children's cognitive, motor, and general neurological development and functioning (Ferguson et al., 2013; Muñoz-Quezada et al., 2013), as well as their physical health (Van Maele-Fabry et al., 2013) and potentially socioemotional functioning, including inattention behaviors and ADHD (Ferguson et al., 2013; Furlong et al., 2014; Hyland et al., 2022). Contact with the clothing of occupationally exposed caregivers is another well documented exposure pathway, although the impacts of these toxins on children are still direct. In this case, all members of the family, including caregivers, siblings, and the target child, are exposed to toxic pesticides from a caregiver's work environment within their home environments. This exposure might well impact both the physical health and attentional behaviors of target children, their siblings, and their caregivers, thus negatively impacting focused attention and subsequent contingent responsiveness in sibling-child and child-caregiver interactions. Further work is needed to evaluate the potential impacts of pesticide exposure on family interactions and processes.

Air Pollution

With ongoing growing urbanization, another key environmental factor directly impacting children and adolescents growing up in the Global North and South is exposure to indoor and outdoor air pollutants, including nitrogen dioxide, sulfur dioxide, carbon monoxide, ozone, and polycyclic aromatic hydrocarbons (Perera, 2017). Such exposures have both short- and long-term impacts on child growth, physical and mental health (including asthma and anxiety), cognitive functioning, attention processes, inhibition, and impulsivity (Bruce et al., 2013; Evans, 2006; Ferguson et al., 2013; Knibbs et al., 2018; Landrigan et al., 2019; Perera, 2017; Sinharoy et al., 2020; Yorifuji et al., 2016).

Beyond the direct effects of air pollution, there is limited work documenting the impacts of air pollutants on family interactions and processes. However, there is good evidence from both cross-sectional and longitudinal studies for the impacts of air pollution on negative affect in adults and children and, under certain circumstances, aggression (Dalton et al., 2020; Evans, 2006). In addition, a recent study following a longitudinal birth cohort in Krakow, Poland, found both direct and indirect impacts of prenatal PAH exposure on child behavioral functioning (Perera et al., 2013). Specifically, psychological distress during the second trimester and prenatal PAH exposure interacted such that higher maternal psychological distress with higher prenatal PAH exposure had a greater impact on child psychological well-being. In addition, some research indicates that, much like heavy metal exposure, caregivers' ability to monitor and control the indoor environments in which their children live can influence the level of exposure to these toxins (Ferguson et al., 2013; Perera, 2017). Thus, children living in families exposed to multiple environmental risk factors are likely to be exposed to greater risk both directly and indirectly, through impacts on caregivers and siblings that then might influence interactions with the target child. Carefully designed environmental systems research is needed to better understand these risks for children, particularly as it is conceivable that caregiver exposure to toxins might in fact alter the caregiver's behavior such that the caregiver is less likely to prevent the child's exposure to these same toxins. Furthermore, the caregiver's resources to help the child cope with the child's toxic exposures may be undermined by the caregiver's own exposure.

Water Pollution, Safety, and Sanitation

Diarrheal dehydration resulting from a lack of access to safe, clean water is still one of the leading causes of illness and death for children under the age of 5 years across the globe (WHO/UNICEF JMP for Water Supply, Sanitation,, & Hygiene, 2021). Yet, 5 years into the Sustainable Development Goals, which include access to safe water, sanitation, and hygiene (WASH), close to 2 billion individuals, most living in the Global South, lack basic access to clean water. Furthermore, many children growing up in the Global South and in low-income communities in the Global North today are exposed to toxins such as lead, arsenic, and tetrachloroethylene (PCE) in the water they have available to drink (Ferguson et al., 2013; WHO/UNICEF JMP for Water Supply, Sanitation,, & Hygiene, 2021). Such exposures have demonstrated impacts on child physical health and motor, cognitive, and socioemotional development and functioning, including some potentially long-term impacts lasting into adulthood (Aschengrau et al., 2011; Bartlett, 2003; Ferguson et al., 2013; Grantham-McGregor et al., 2007; Landrigan et al., 2019). Further, child exposure to bacteria-contaminated water results in diarrhea and intestinal parasites that contribute to malnutrition and stunting (Bartlett, 2003; Mshida et al., 2018). Malnutrition and stunting often also result in hyporesponsive infants who offer less positive reinforcement for sensitive and responsive parenting.

Beyond these direct effects, a lack of access to clean water is associated with school absences due to illness and the need for girls to walk long distances to collect clean water (Bartlett, 2003). This necessary time commitment on the part of children and their families to securing drinking water may well impact family routines and rituals, as well as other family processes, but there is little known work studying these types of disruptions in isolation. Indeed, families that lack access to clean water tend to be exposed to multiple environmental risk factors. The potential impacts of a lack of access to safe water and sanitation on family life are important to understand, given the scope of the problem. Furthermore, it is important to note that the individuals in the household most disrupted by the need to walk long distances for water are older girls and their mothers, as well as younger children who need to be cared for by less competent younger female siblings when their mothers or older sisters are fetching water. This might result in increased risk of injury for young children and less time available for positive caregiver-child verbal and play interactions. Primary caregivers and older siblings may also have reduced physical and cognitive energy to engage in positive interactions with younger children following extensive time and energy investments in water collection.

As discussed throughout this section, while the direct short- and long-term impacts of toxins and pollutants on child health and development across the globe have been extensively studied in the last two decades, there is very little work investigating influences on family interactions and processes. Yet, we know that vulnerable children and families are disproportionately exposed to multiple toxins and pollutants in their home, school, and work environments and that these exposures have long-term, sustained impacts on child, adolescent, and adult health and well-being (Carter-Pokras et al., 2007; Evans, 2003, 2006; Ferguson et al., 2013; Perera, 2017). It is thus likely that such exposures impact family interactions and processes and that mitigation strategies and other methods of reducing or controlling exposure to these toxins interrupt and disrupt key family routines and rituals. Potential impacts include reduced patience and empathy on the part of both caregivers and siblings, reduced contingent responsiveness, and reduced ability to sustain and scaffold positive caregiver-child and sibling-child interactions over time, all of which can in turn potentially impact the development of self-regulation and a secure attachment relationship. In addition, children's cognitive development and functioning are likely impacted by reduced time available for and sustained attention given to cognitive enrichment activities such as reading, informal learning, music, and art. Further, caregivers likely have more limited time and capacity for planning and organizing activities, routines, and rituals. Finally, caregiver concerns over potential child exposure to toxins and pollutants in their home and neighborhood environments likely lead to the introduction of constraints on child play. Clearly, further work in this area is warranted.

Temperature and Lighting Approximately 5% of adults, some adolescents, and a small number of children who are exposed to shorter daylight hours suffer from seasonal affective disorder (SAD), a form of depression characterized by increased sadness, fatigue, and depression during the winter season (Booker & Hellekson, 1992; Rohan et al., 2009; Rosen et al., 1990; Swedo et al., 1995; Tonetti et al., 2007). Depression impacts mood and cooperative social behavior both in the home and in school settings (Evans, 2003; Küller & Lindsten, 1992; Rohan et al., 2009; Tonetti et al., 2007). Further work to better understand the subsequent impacts on family interactions and processes is warranted, particularly as caregiver and sibling depression likely negatively impacts their social engagement with and responsiveness to the target child. Depression experienced by the target child is likely to rend them less responsive and engaged and less willing to seek out social interactions within their family environments. The resultant reduced and poorer quality sibling-child and caregiver-child interactions likely negatively impact child and family members' well-being and long-term socioemotional functioning. Furthermore, disrupted sleep cycles and depression are both risk factors for poor quality sleep. Poor quality sleep reduces the energy caregivers, siblings, and target children have to engage in positive interactions with family members, interferes with patience, and makes it more challenging for family members to be attuned to the needs of those around them.

Some evidence in school and work environments in particular suggests impacts of temperature and climate control on child and adult functioning and learning, including mood, sociability, and school achievement (de Dear et al., 2015; Earthman, 2004; Evans, 2006; Uline & Tschannen-Moran, 2008). Poor climate control also contributes to asthma and related health conditions, as well as the likelihood of children being sick in school and absent from school as a result of illness (Jackson et al., 2013). Poor climate control within home environments, although less studied, likely has similar impacts. In addition, poor climate control and a lack of adequate temperature control coupled with indoor air pollutants within home environments have a cumulative impact on child, sibling, and caregiver physical and mental health and well-being, including quality sleep (see also Evans, 2003, 2006). We also know that air pollution compounds temperature control challenges in indoor spaces and that household energy insecurity is associated with household food insecurity, child health, and psychological adjustment (Cook et al., 2004). Further work investigating the subsequent impacts on family interactions and processes is certainly warranted, given the likely negative impacts of poor climate control coupled with food insecurity and energy insecurity on caregiver, sibling, and target child psychological distress and stress. The increased unpredictability and stress within home environments with poor climate control, coupled with increased irritability, might well lead to higher levels of conflict and the disruption of positive, sustaining family routines and rituals.

Noise The direct impacts of noise on individual children, siblings, caregivers, and teachers and the subsequent impacts on sibling-child, child-caregiver, and child-teacher interactions have been far better studied than the impact of other physical environmental factors discussed thus far (for reviews, see Evans, 2003, 2006; Doan & Evans, 2020; Ferguson et al., 2013; Whipple & Evans, 2022). Ambient noise (most often from transportation sources such as road traffic, public transit, or airplanes) impacts infant, child, and adult executive functioning, short-term memory, auditory discrimination, reading, hyperactivity, inattention, motivation, task persistence, perceived stress, and blood pressure. It also impacts annoyance, aggression, hostility, child academic achievement, behavioral conduct disorders, social adaptability, and oppositional behaviors at home and at school. Although research on the impacts of noise on child and adult functioning in the Global South is more limited, findings to date demonstrate similar effects. These impacts are a result of both the aversive nature of chronic noise and its uncontrollability and unpredictability (Evans, 2006). Repeated exposures to uncontrollable and/or unpredictable events such as noise are stressful and can undermine motivation and task persistence, thus resulting in learned helplessness behaviors (Cohen et al., 1986; Evans & Stecker, 2004; Hiroto, 1974).

Beyond the direct impacts of ambient noise in homes, schools, and neighborhood settings, noise indirectly impacts child functioning through altered caregiving and teaching behaviors and sibling interactions (Doan & Evans, 2020; Evans, 2006; Ferguson et al., 2013; Whipple & Evans, 2022). In high noise schools, teachers report fatigue, annoyance, decreased outdoor activities, interruptions of teaching, and alterations in teaching methods to adjust for noise levels (Evans, 2006; Klatte et al., 2017). In high noise homes, caregivers are less attentive and responsive, parents tend to engage in less verbal interaction with their children, and children tend to engage in less social interaction (Fiese et al., 2015; Kirkorian et al., 2009; Matheny Jr. et al., 1995; Pempek & Lauricella, 2017). Fatigue, physiological stress, lower parent efficacy, and direct interruptions of loud noise may lead to parents and teachers talking and reading less to children and being generally less responsive (see also Corapci & Wachs, 2002). These caregiver behaviors, combined with greater social withdrawal, poorer social skills, and decreased motivation and task persistence on the part of target children and their siblings, likely have cumulative impacts on the health and well-being of everyone in the household.

Noise can impact family interactions and processes by reducing interactions, by producing negative interactions, and by disrupting key family routines and rituals. In a related experimental study of the impacts of background television on caregiver-child interactions, parents of 12-, 24-, and 36-month-old children were less responsive when a television program was playing in the background (Kirkorian et al., 2009). The quality of caregiver-child interactions was also reduced, even though overall child responsiveness was not impacted. This is particularly concerning given the positive impacts of active caregiver contingent responsiveness and engagement on the complexity of toddler play, as well as the length of time toddlers are able to sustain play episodes.

In another experimental study, social interactions and communications during a family mealtime of families with at least one child aged 5 through 13 years were compared as a function of typical background noise or with continuous loud noise from an adjacent room (Fiese et al., 2015). Families exposed to loud background noise were more frequently distracted and engaged in fewer positive social interactions. This work highlights the importance of understanding the role that noise and other forms of chaos play in disrupting valuable family routines and rituals.

Crowding As with noise, there is a significant body of evidence to date documenting the direct influences of household crowding (typically measured as number of people per room) on child, adolescent, and adult physical and mental health and motor, cognitive, and socioemotional development and functioning across the Global North and South (Booth & Edwards, 1976; Doan & Evans, 2020; Edwards et al., 1994; Evans, 2006; Ferguson et al., 2013; Saegert, 1982; Wachs & Corapci, 2003; Whipple & Evans, 2022). For children, crowding impacts psychological distress, blood pressure, infant psychomotor development, school achievement and IQ, time on task, joint attention, vocabulary, aggressive behaviors, behavioral adjustment at school, learned helplessness behaviors, and social withdrawal. In considering these impacts on children and their families, it is also important to acknowledge that low-income and vulnerable children living in the Global North and the majority of children living in the Global South are more likely to live in crowded home environments (Britto et al., 2017; Evans et al., 1998; Ferguson et al., 2013; Jones et al., 2017).

Beyond the direct effects of crowding on children and adolescents, crowded conditions result in psychological distress as a result of unwanted social interactions, resulting in social withdrawal on the part of target children, siblings, and caregivers within home environments and target children, peers, and teachers within school environments (see also Doan & Evans, 2020; Evans, 2006; Ferguson et al., 2013; Wachs & Camli, 1991; Whipple & Evans, 2022). This withdrawal disrupts socially supportive sibling-child, caregiver-child, peer-child, and teacher-child relationships. In addition, caregivers demonstrate lower levels of child stimulation, responsiveness, and verbal interaction when operating within the same crowded spaces as their children (Evans et al., 1999; Wachs & Camli, 1991). For example, in a secondary data analysis of a longitudinal study, relationships between crowding, family socioeconomic status, and parents' language diversity and verbal responsiveness to a target child aged 6–36 months were investigated (Evans et al., 1999). Crowding significantly predicted language diversity and parental verbal responsiveness, controlling for socioeconomic status. In addition, parental verbal responsiveness mediated the relationship between crowding and language diversity. These findings together suggest that, in more crowded homes, parents speak in less complex ways to their children, partly as a result of reduced verbal responsiveness to their children's behavior. This study thus demonstrates clear impacts of crowding on caregiver-child verbal interactions.

Crowding can disrupt children's exploration and play, as well as their general engagement with both objects and people in their immediate environments (Ferguson et al., 2013; Heft, 1979; Liddell & Kruger, 1987, 1989). In addition, work with rodents shows long-term impacts of social crowding of pregnant females on pup birthweights and even later timing of puberty and reproductive behaviors, in addition to impacting parental responsiveness in the short term (Beery & Kaufer, 2015; see also Calhoun, 1962; Christian & LeMunyan, 1958; Harvey & Chevins, 1987). A recent investigation of 9-, 15-, and 36-month-old children living in the United States and the United Kingdom tested the potential mediating role of maternal responsiveness as a partial mediator of the relationship between residential crowding and child cognitive development (Evans et al., 2010). In the first study, 15- and 36-month-old infants living in higher residential density conditions had compromised current and future cognitive development. Further, when maternal responsiveness was added in the equation, cognitive development at 36 months was no longer predicted by either current (36-month-old) or past (15-month-old) residential density. In the second study of 9- and 36-month-old infants, although residential density at 36 months predicted child cognitive development, both with and without maternal responsiveness included in the model, adding maternal responsiveness significantly reduced the predictive power of residential density alone. Taken together, these studies demonstrate strong evidence for the mediating role of maternal responsiveness in explaining the well-documented relationship between residential crowding and child cognitive development.

While there is clear evidence for the impacts of crowding on social interactions within the family and for the impact of social interactions within the family on child health and well-being, only a small number of studies have directly tested this complete system. One interesting example comes from studying 10- to 12-year-olds and their families living in very crowded conditions in urban India (Evans et al., 1998). Household crowding was associated with behavioral adjustment at school, family conflict, lower levels of family support, learned helplessness for girls, and elevated resting blood pressure for boys. Importantly, these relationships were partially mediated by caregiver-child conflict. In this case, crowding may well have impacted both the parent and the child directly, thus influencing their interactions with each other and so indirectly impacting child health and well-being.

Another quasi-experimental study investigated associations between the level of crowding in participants' home environments (people per room) and participant likelihood to seek support from a confederate and rate the confederate as supportive in stressful laboratory conditions (Evans & Lepore, 1993). In this way, researchers directly tested the hypothesis that people living in more crowded environments are more likely to socially withdraw and less likely to have positive social interactions. Indeed, people from more crowded households were less likely to seek social support, perceived that less social support was offered to them, and offered less social support when exposed to an individual in need of social support. The findings highlight the impacts of crowding on social interactions within the home environment and provide an explanatory mechanism for the documented impacts of crowding on family interactions. They also highlight Bronfenbrenner's insights regarding cross-context influences of environments. Home crowding influenced behavior in the laboratory. With growing urbanization and a high percentage of low-income and vulnerable children in the Global North and South spending much of their lives in crowded homes and schools, the importance of better understanding both the direct and indirect impacts of crowding on child health, well-being, and development is clear. Given the evidence thus far for key impacts of residential and school crowding on family interactions and processes, further work in this area is clearly warranted. This is an area where further thinking about the role of design in crowding and other environmental stressors may also yield potentially useful amelioration strategies where the removal of the stressor (e.g., not enough rooms for people) is not feasible. For instance, some work indicates that brighter light, views of nature, and opportunities to at least momentarily remove oneself from social interaction reduce some of the adverse impacts of crowding (Rollings & Evans, 2019).

Chaos and Instability Noise and crowding are two aspects of environmental chaos. Other frequently studied components of chaos in home, school, and child-care environments include cleanliness, clutter, home traffic (frequency and predictability of movement of people in and out of the residence), inconsistency, unpredictability, disorganization, and a general lack of routines (see Doan & Evans, 2020; Evans & Wachs, 2010; Whipple & Evans, 2022). Food and health insurance insecurity can also contribute to household chaos. Settings that are unpredictable and unstructured might include few routines, and those routines that exist are likely to be interrupted and disrupted on a regular basis (Evans & Wachs, 2010; Weisner, 2010). This facet of chaos likely plays a key role in the negative impacts of chaos on child and family health and well-being, as chaotic environments interfere with effective proximal processes (Bronfenbrenner & Evans, 2000; Evans & Wachs, 2010; Fiese, 2006). Chaos negatively impacts child, adolescent, and adult physical and mental health and motor, cognitive, and socioemotional development and functioning (Andrews et al., 2021; Doan & Evans, 2020; Evans & Wachs, 2010; Ferguson et al., 2013; Fiese & Winter, 2010; Marsh et al., 2020; Wachs & Corapci, 2003; Whipple & Evans, 2022). Specific direct impacts on children and adolescents include poorer executive function, language development, self-regulation, behavioral control, and ability to understand social cues and higher levels of psychological distress, learned helplessness, depression, anxiety, aggression, and internalizing and externalizing behavioral problems.

The majority of work to date on both direct and indirect impacts of chaos on child health and well-being has been conducted in the Global North (Ferguson et al., 2013; Weisner, 2010). However, the extant research to date in the Global South documents similar impacts of chaos on health and socioemotional development and functioning (Akram & Shamama-tus-Sabah, 2020; Shamama-tus-Sabah et al., 2011; Shamama-tus-Sabah & Gilani, 2010; Wachs & Corapci, 2003; Weisner, 2010). Relatedly, although research on the impacts of residential mobility on children growing up in the Global South is quite limited, the work to date documents negative cognitive and socioemotional impacts on children and families (Bartlett et al., 1999; Dizon & Quijano, 1997; see also Bures, 2003; South et al., 2005). Residential mobility in the Global North has demonstrated impacts on academic achievement, psychological adjustment, and socially supportive peer relationships

(Jelleyman & Spencer, 2008; Oishi, 2010). Given high levels of residential mobility among low-income families living in urban areas in the Global South (Bartlett et al., 1999; Chatterjee, 2007), further work is clearly warranted. This being said, in both defining chaos and evaluating its impacts on child and family well-being, it is essential to consider potential variations in individual and family perceptions and experiences of chaos within various cultural contexts (Weisner, 2010). It is thus most important to consider the ways in which chaotic settings interfere with the sustenance of family routines and rituals that contribute to individual and family well-being in various contexts across the globe.

Given the nature of household chaos, a reasonable body of research over the past two decades has directly evaluated its impacts on family interactions and processes, with the majority of this work focusing on parenting in chaotic environments. Household chaos is associated with increased caregiver-child conflict, punitive and harsh parenting behaviors, and paternal hostility towards the child. Household chaos is associated with decreased caregiver-child closeness, supportive authoritative parenting, parental emotional availability and support, parental responsiveness, parental promotion of child exploration and play, parental stimulation (linguistic and object-directed), and parental efficacy (for recent reviews, see Ackerman & Brown, 2010; Doan & Evans, 2020; Evans & Wachs, 2010; Marsh et al., 2020; Whipple & Evans, 2022). However, less work has specifically studied the impacts of household chaos on family processes, including impacts on the construction of family time (including routines and rituals), the disruption of family activities (including routines), and family meaning making (including both the establishment and maintenance of family rituals and the meanings created out of disruption and unpredictability; Fiese et al., 2006; Fiese & Winter, 2010). Further work to investigate the specific mechanisms by which chaos impacts family interactions and processes, and subsequent child health and well-being consequences, is warranted, particularly in the Global South.

Several recent studies have directly tested the mediating and/or moderating role of parenting behaviors and other components of family interactions in the relationships between household chaos and child health and well-being. The interrelationships between first grade (6 or 7 years of age) child conduct problems and callous-unemotional behaviors; parenting behaviors at 2, 6, 15, 24, and 36 months of age; and household chaos (disorganization and instability) were assessed at the same time points among non-urban lower-income US families (Mills-Koonce et al., 2016). Associations between child conduct problems and callous-unemotional behaviors and both family socioeconomic status (as measured by family income and parental education) and household chaos were mediated by parenting behaviors, specifically maternal sensitive behaviors and harsh-intrusive parenting behaviors.

In another study of parenting behaviors, household chaos, and early child development using the same sample, researchers evaluated whether household chaos, as measured at 2, 6, 15, 24, and 36 months of age, predicted behavioral regulation in kindergarten and whether parent responsivity and acceptance during the first 3 years of life mediated these relationships (Vernon-Feagans et al., 2016). Researchers also investigated whether parenting might be a partial mediator between household chaos and child executive function at 3 and 5 years of age. One component of household chaos, disorganization (but not household instability), during the first 3 years of life indirectly impacted child executive functioning at 3 and 5 years of age through direct impacts on parenting responsivity and acceptance. These associations remained beyond the impacts of family socioeconomic status. A body of controlled experimental research with rodents and non-human primates similarly demonstrates the negative impacts of chaotic environments (created through variations in foraging demand or food availability) on parent (maternal) behaviors, specifically maternal contingent responsiveness, as well as subsequent impacts on their offspring's behavior (see, e.g., Coplan et al., 2017). These variations in both parent and child behavior are accompanied by parallel changes in the functioning of the HPA axis, in other words, both short- and long-term changes in the biological response to stress (Coplan et al., 2017). Further work with humans is warranted to better understand the specific mechanisms underlying the impacts of household chaos on family interactions and processes, and specifically short- and long-term impacts on parent, sibling, and target child stress systems.

Beyond residential chaos, housing instability and subsequent residential mobility interfere with social support within home environments (Marcal, 2021). Importantly, a change in household residence is frequently accompanied by a change in school. In one investigation of the relationships between housing security and parenting behaviors, researchers tested a series of direct and indirect pathways between housing insecurity and adverse parenting behaviors towards adolescents, considering the potential mediating roles of parenting stress and maternal depression (Marcal, 2021). Housing insecurity was significantly associated with increased parenting stress and maternal depression, and parenting stress – but not maternal depression – was significantly associated with adverse parenting behaviors. Finally, housing insecurity was not directly associated with adverse parenting. Thus, parenting stress mediated the relationship between housing insecurity and maternal adverse parenting behaviors.

In future work investigating the impacts of residential chaos, housing instability, and residential stability, it is important to consider variations in residential and caregiving models across global contexts. For example, in much of the Global South, children have multiple primary caretakers, potentially living in multiple households, across both shorter (weeks) and longer (months, years) time spans (Weisner, 2010). In such contexts, careful consideration must be given to defining relevant caregivers, siblings, and residential contexts for a target child. The same may be true for many families in the Global North as well, particularly but not exclusively in migrant and immigrant communities. Indeed, in such communities, children may move multiple times across regions and countries over the span of months and years. Defining relevant residential contexts at different points across time and development is thus essential if we are to understand the lived experiences of global children and their families.

Housing Quality Housing quality, including structural quality; maintenance; the availability of amenities such as running water, electricity, and heating; and the existence or absence of physical hazards, has demonstrated impacts on child, adolescent, and adult physical and mental health and well-being (Cook et al., 2004; Evans, 2003, 2006; Evans et al., 2003; Ferguson et al., 2013; Jones et al., 2017; Wells & Harris, 2007). Importantly, though, much of this work has not disentangled the direct impacts of individual components of the home environment, including both physical and psychosocial components. More extensive work in the Global South is also warranted, particularly given the greater degree of heterogeneity in housing quality throughout many lower-income countries. A recent eight-site study of mothers and infants in Bangladesh, India, Nepal, Pakistan, Brazil, Peru, South Africa, and Tanzania documented both the usefulness of adapting the HOME scale (Bradley et al., 2003; Bradley & Caldwell, 1984) for use across diverse contexts in the Global South and the possibility of evaluating the impacts of specific components of the HOME scale (including identifying a subset of items focused on key aspects of the physical environment; Jones et al., 2017). In addition, some limited work in the Global South documents the potential impacts of the availability of learning materials and resources such as writing materials and books, as well as electricity, running water, a radio, television, and a transportation vehicle, on children's cognitive functioning and play (Ferguson, 2008; Gauvain & Munroe, 2009; Hamadani et al., 2010). Clearly further work disentangling the impacts of specific components of the physical home environment on child health and well-being is warranted.

Beyond the direct impacts of housing quality and the availability of resources within the home environment, housing quality can remove or attenuate the function of home as a restorative environment for children and their families. Chaos within the home environment, coupled with exposure to indoor toxins and pollutants, further attenuates the ability of home to function as a positive, restorative space. Although direct work assessing these interrelationships is needed, there is reasonable evidence that adults living in poor quality housing have higher levels of psychological distress, with a limited number of longitudinal studies indicating that these impacts are directional (Evans et al., 2001; Wells & Harris, 2007; Wells & Moch, 2003). There is also some evidence that poor quality housing results in social withdrawal (Wells & Harris, 2007). Research documenting the impacts of housing type (especially high-rise dwelling) similarly suggests that social withdrawal and parental limitations of child play negatively impact child cognitive development and socioemotional functioning (Evans, 2003, 2006). Given the known impacts of adult psychological distress and social withdrawal on child health and well-being, it is certainly conceivable that housing quality would impact family interactions and processes. Furthermore, housing quality impacts parents' restriction and control of children's play environments, thus again likely impacting child health and development (Ferguson et al., 2013; Gauvain & Munroe, 2009). Furthermore, it is likely that the constant challenge of dealing with a poor quality residential environment, including exposure to toxins and hazards, contributes to parental anxiety and worry, thus further impacting caregivers' time and ability to engage in positive interactions with target children, their siblings, and other adult members of the household. Clearly further work in this area is warranted.

Extensions of Home

Beyond the impacts of the immediate home environment on child and family health and well-being, we discuss the neighborhood, play spaces, afterschool programs, and youth centers as extensions of home in which children spend a significant percentage of time on a regular basis. For infants and young children, play spaces including parks and playgrounds and other safe, potentially green play spaces are particularly important to consider. It is also worth noting that children are most likely to occupy these spaces in the company of a caregiver and/or potentially one or more siblings. For elementary and middle school children, parks, playgrounds, and other play spaces continue to be important, although at these ages children may occupy the spaces without caregivers and/or siblings. Afterschool settings become more important at this age. Finally, in adolescence, multiple neighborhood spaces may become salient, with youth centers being particularly important indoor spaces. These spaces are typically not simultaneously occupied by caregivers, although of course other adults and potentially older siblings may be present. In this section, we discuss some of what is known of the impacts of these settings on child health and well-being, with a focus on impacts on family interactions and processes when possible.

Neighborhood Neighborhood quality is associated with child, adolescent, and adult health and well-being and child cognitive and socioemotional development (Christian et al., 2015; Diez Roux & Mair, 2010; Evans, 2003, 2006; Ferguson et al., 2013; Leventhal & Brooks-Gunn, 2000). However, typically, measures of neighborhood quality focus on social factors, principally SES, with scant attention paid to impacts of specific neighborhood physical features such as traffic, housing dilapidation, vacancies, and access to natural features. Because neighborhood quality is strongly associated with inhabitants' SES, the direct effects of key components of neighborhood physical quality on child health and well-being are unclear. Yet, poor quality neighborhoods, those with high levels of air and water pollution, poor sanitation, high housing density, high noise, high traffic, limited street lighting, and limited access to open green spaces, safe play spaces, and grocery stores, are common for BIPOC children in the Global North and for the majority of children growing up in the Global South (Bartlett, 1999; Chawla, 2002, 2015; Evans, 2004, 2006; Ferguson et al., 2013; Hardoy et al., 2001; Kruger & Chawla, 2002; Leventhal & Brooks-Gunn, 2000). Thus, examining the unique influences of key aspects of the physical neighborhood environment on child, adolescent, and adult health and well-being is essential to better understanding the contexts in which children and families live their lives (Villanueva et al., 2016).

A small number of studies have demonstrated direct impacts of poorer physical quality neighborhoods on child health and development, independent of family SES. For example, 9- to 11-year-olds living in poorer physical quality neighborhoods had higher levels of psychological distress, independent of family SES (Homel & Burns, 1989). In this work, however, individual measures of key aspects of the physical environment were not available. Some limited work has considered proximity to street traffic and found that this variable is independently correlated with pediatric injury risk, reduced interactions with neighbors, restrictions in outdoor play, and smaller social networks for young children (Aarts et al., 2012; Appleyard & Lintell, 1972; Hüttenmoser, 1995; Macpherson et al., 1998; see also Christian et al., 2015 for a recent review). In addition, it is known that the general safety and walkability of neighborhoods impact child and adult physical activity and social interactions and thus their cognitive and socioemotional functioning (Alparone & Pacilli, 2012; Carlson et al., 2015; Giles-Corti et al., 2009; Villanueva et al., 2016). Child physical activity is partly impacted by parental restrictions on play resulting from fear of crime within the neighborhood.

In considering the impacts of the neighborhood physical environment on family interactions and processes, interestingly, the most expansive work to date on the impacts of the neighborhood physical environment on child health and development explicitly assesses relationships between parents' perceptions of neighborhood safety and children's socioemotional development and health (Christian et al., 2015; Edwards & Bromfield, 2009; Evans, 2006; Fan & Chen, 2012; Ferguson et al., 2013). It should be noted, however, that to date this work looks primarily at associations between variables and does not assess the underlying mechanisms behind these associations.

In a large study of 4983 Australian children aged 4 to 5 years living in 257 neighborhoods, children's conduct problems were associated with four dimensions of the neighborhood: neighborhood socioeconomic status, neighborhood safety, neighborhood cleanliness, and neighborhood belonging (parents' trust of neighbors, a sense of identity with the neighborhood, how well informed they were about local affairs, and knowledge about where to find information about local services; Edwards & Bromfield, 2009). Additionally, relationships between neighborhood socioeconomic status and children's conduct problems were mediated by parent perceptions of neighborhood safety and neighborhood belonging; associations between neighborhood safety and conduct problems were mediated by neighborhood belonging. Child pro-social behavior was directly associated with parent perceptions of neighborhood cleanliness and neighborhood belonging.

In a study of factors impacting fifth grade children's walking to school in various neighborhoods, parents and children typically agreed on walking barriers. Although in less walkable communities, parents perceived more walking barriers than their children did (Napier et al., 2011). The frequency of walking to school was impacted by both child and parent perceptions of walking barriers and proximity to school. Importantly, walking to school had positive impacts on child BMI.

Beyond parents' perceptions of neighborhood environments, in UNESCO's Growing Up in Cities project (Chawla, 2002), 10- to 15-year-olds growing up in

South Africa (Kruger, 2002), India (Bannerjee & Driskell, 2002), and Argentina (Cosco & Moore, 2002) identified a number of key aspects of their neighborhood environments that impacted their play and other peer interactions, including high traffic, litter, poor sanitation, and a lack of open green spaces. Resource availability becomes important for adolescents as they look for nearby job, internship, and other training and networking opportunities as part of their continued education and professional development. Access to safe, green play spaces is important for peer interactions across the lifespan. We discuss play spaces, afterschool programs, and youth centers in more depth below.

Play Spaces Safe accessible play spaces, including parks, playgrounds, public school yards, and other areas within children's neighborhoods are important settings for child play and other caregiver-child, sibling-child, and peer interactions. Children growing up in the Global North and South indicate a preference for and engage in more complex play in safe, green play spaces (Bartlett et al., 1999; Evans, 2006; Ferguson et al., 2013; Gill, 2014; Loebach & Cox, 2020; Maxwell et al., 2008; van Dijk-Wesselius et al., 2018). The most positive of these types of spaces include spaces with places to sit, spaces where children can gather away from adult monitoring, spaces that provide safety and distancing from traffic and noise, spaces with natural elements, and spaces with physical characteristics and materials conducive to play, particularly when there is some gradient of opportunity to test out physical and cognitive competencies.

Natural settings positively impact both children and adults by serving a restorative function in the face of chronic stress and cognitive fatigue (Kaplan & Kaplan, 1989; Kaplan & Talbot, 1983) while also providing opportunities for positive social engagement and more complex forms of motoric and social play (Evans, 2006; Faber Taylor et al., 1998; Fjørtoft, 2004; Heft, 1988; Kyttä, 2002, 2004). Safe green play spaces have positive impacts on children's executive functioning, including attentional processes and emotional self-regulation (Faber Taylor et al., 2002; Kuo & Faber Taylor, 2004; Wells, 2000), academic achievement (Bell & Dyment, 2008; Evans, 2006; van Dijk-Wesselius et al., 2018), general well-being (Hattie et al., 1997), and responses to stressful life events (Wells & Evans, 2003). Children also engage in greater levels of physical activity in safe green play spaces, thus positively impacting physical and mental health (Evans et al., 2010; Evans et al., 2012a, b).

There is limited direct evidence for the impacts of safe, accessible play spaces on family interactions and processes, although it is important to note that these are common places of sibling-child, peer-child, and, for younger children, caregiver-child interaction. Thus, the positive impacts on children documented here should apply to all children and adults interacting within these spaces. Additional work with adults documents positive impacts of natural environments on collective efficacy (Cohen et al., 2008), adult physiological stress responses (see Evans, 2003), and cognitive fatigue (Evans, 2003, 2006), all with known positive impacts on adult mood, well-being, social engagement, and responsiveness. Furthermore, although little work has directly studied child-child interactions in play settings, particularly in light of the impacts of child health and well-being, it is likely that these interactions play a key role in children's experiences of all types of play spaces (Ferguson et al., 2017; Loebach & Cox, 2020).

A number of recent studies have sought to better understand parent perceptions of neighborhood play and related spaces and subsequent impacts on children's engagement in these spaces (Villanueva et al., 2016). Although these studies do not directly assess family interactions or processes, they provide insight into potential influences on family interactions within child play spaces. For example, in a study of relationships between the physical activity of 6- to 11-year-old children and a caregiver's perceptions of their neighborhood environments, children's physical activity was correlated with parent perceptions of proximity of play areas (to home) and street connectivity (Tappe et al., 2013). Parents' perceptions of neighborhood safety and the availability of walking and cycling facilities were positively associated with parent-reported child activity levels in public spaces such as parks. Importantly, within this age range, children may have limited opportunities to access public play and activity spaces within their neighborhoods without the presence of a caregiver or possibly an older sibling. Thus, parents' perceptions and availability both impact children's opportunities for play and engagement in natural and other public play spaces within their neighborhood environments (see also Loebach et al., 2021).

Afterschool Programs and Youth Centers Afterschool programs and youth centers (including identity spaces, community spaces, and spaces for community action, internships, and work and college preparation) are an important extension of home environments for children in elementary through high school. Importantly, given the amount of time children might spend in these settings, the quality of the physical environment (toxins and pollutants, pesticides, temperature and lighting, noise, crowding, chaos, structural quality, presence or absence of safe, potentially green play and interaction spaces) in these spaces likely has similar impacts as those documented above for home environments.

Afterschool programs and youth centers are important spaces for peer-child and, potentially, sibling-child interactions. Although primary caregivers are unlikely to spend much time in these spaces, there is also potential for child-adult interactions beyond those with children's primary caregivers. Documented impacts of the physical environment on children and adults are thus likely to also impact children indirectly through their impacts on other children and adults operating within these settings on a regular basis.

Youth intervention work led by adults in collaboration with youth frequently focuses on place making, community action work, and youth identity within and beyond structured youth centers (see, e.g., Chawla, 2002; Jo et al., 2018; Kruger, 2002; Loebach et al., 2020; Prince, 2014). Work in this area is still limited, but has significant potential for better understanding adolescent health and well-being. Particular areas of interest might include further exploration of the balance between adolescent needs for autonomy and identity development apart from parents along with needs for supportive monitoring and collaborative self-regulation.

A recent relevant study employed a place-based approach to investigate the urban experiences of 1341 Finnish and Japanese fifth and eighth grade children residing in Helsinki and in Tokyo (Kyttä et al., 2018). As expected, in both contexts, children's independent mobility within urban spaces outside of direct adult supervision was highly relevant to them. Beyond independent mobility, however, affordances for safe play, time in nature, movement, exploration, and time meeting friends were highlighted across contexts. Children also highlighted negative affordances of some less child-friendly urban spaces, including areas they identified as dangerous, dirty, noisy, hectic and crowded, and boring; places where children were not welcomed; and, importantly in the context of the current chapter, places parents did not allow them to be. High traffic areas received particularly negative ratings. Clearly, further collaborative work with children across global contexts to better understand the direct and indirect impacts of varying physical environments on their health and well-being would be valuable.

Beyond Home

Beyond immediate home environments, two additional settings that have significant direct and/or indirect impacts on child health and well-being through their impacts on family interactions and processes are target child and sibling school and childcare settings and caregiver and sibling work settings. Many of the physical environment characteristics discussed thus far, including toxins and pollutants, pesticides, temperature and lighting, noise, crowding, chaos, and structural quality, have documented impacts on child and adult health and well-being. In addition, we discuss the limited work to date on the impacts of these settings on family interactions and processes.

School and Childcare As is the case for afterschool programs and youth centers, children spend a significant portion of their time in school and childcare facilities. Thus, the same characteristics of the home physical environment likely have similar direct and indirect impacts on children through their impacts on siblings, peers, and caregiving adults operating within these environments (for reviews, see Evans, 2006; Ferguson et al., 2013; Glewwe et al., 2013; Higgins et al., 2005; Irwin et al., 2007; Riddell, 2008). Research in childcare and school settings shows direct impacts of temperature control (Cash, 1993; Earthman, 2004), air quality (Chithra & Madanayak, 2018; Earthman, 2004), natural lighting (Heschong, 1999), noise (Klatte et al., 2017; Werner et al., 2015), and class size (Ehrenberg et al., 2001; Greenwald et al., 1996) on academic achievement; direct impacts of childcare crowding on attentional processes (Maxwell, 1996); and direct impacts of classroom crowding on child off-task time (Krantz, 1974), attention, aggression, and other classroom behaviors (Evans, 2006; Grantham-McGregor et al., 1998).

There is also a reasonable body of research documenting the impacts of poor quality school buildings on teacher attitudes, behaviors, and performance (Buckley et al., 2004; Uline & Tschannen-Moran, 2008), as well as on teacher retention and attrition (Buckley et al., 2004). The physical environment factors measured include air quality, temperature, natural light, and noise. These factors impact teachers' health, perceived health, and physiological stress levels (Buckley et al., 2004), as well as their responsiveness to children. Given the known impacts of teacher responsiveness, commitment, and collegiality on student achievement (e.g., Hoy & Sabo, 1997), it is likely that poor quality school environments negatively impact relevant teacher-child interactions that in turn negatively impact child health and well-being. Teacher turnover is higher in poor quality school environments, and teacher attitudes towards their job are more negative. Both factors likely have impacts on child-teacher interactions and subsequent impacts on child health and well-being.

In considering the impacts of the physical environment of schools and childcare facilities on target children, siblings, and adult teachers and caregivers, it is important to note that a significant percentage of school buildings and grounds across the Global North and South are in disrepair (Glewwe et al., 2013; Riddell, 2008; UNICEF, 2010; von Ahlefeld, 2007). Rural schools in the Global South in particular are unlikely to have finished flooring, electricity, water, and basic sanitation facilities (UNICEF, 2010). Clearly, such environments are not conducive to effective teaching, learning, and social interactions between children and between children and their teachers. Furthermore, climate change-related natural disasters such as extreme storms and flooding often interrupt school attendance, and in the Global South, such residential and school displacements may be permanent (Bartlett, 2008). Thus, further work investigating specific direct and indirect impacts of key components of childcare and school physical environments on children is needed.

The impact of redesigning (greening) a rural Austrian middle school schoolyard was investigated using a quasi-experimental design (Kelz et al., 2015). The purpose was to investigate the impact on middle school children's pre- and post-redesign executive functioning, psychological well-being, physiological stress, and perceptions of the schoolyard as restorative. Comparisons of child measures pre- and post-redesign were made for children attending control schools that did not undergo a redesign. Clear positive impacts of the schoolyard redesign were found on physiological stress and psychological well-being, but not on executive functioning. Children's perceptions of the restorative value of the schoolyard also increased.

A longitudinal intervention study in the Netherlands assessed both direct and indirect impacts of greening schoolyards on 7- to 11-year-old children's appreciation for the play space, physical activity, attentional restoration after recess, socio-emotional well-being, and social orientation towards peers (Van Dijk-Wesselius et al., 2018). Both individual and peer interaction impacts were found. Specifically, greening schoolyards positively impacted children's appreciation for the play space, increased physical activity for girls, and improved performance on attentional tasks post-recess. In addition, children reported higher levels of social support and fewer peer problems. Younger children had a more positive prosocial orientation immediately after recess in green spaces. These findings demonstrate the potential impacts of access to green play spaces in school environments on peer-child interactions and suggest potential impacts of similar environments beyond school settings on sibling-child interactions as well.

Another study investigated the factors underlying the impacts of the school physical environment on children's school achievement, specifically associations between middle school teachers' ratings of the quality of school facilities, resource support, and school climate and student academic achievement (Uline & Tschannen-Moran, 2008). The quality of school facilities was associated with student achievement. Importantly, they also found that school climate (e.g., teacher professionalism, community engagement) mediated the relationship between the quality of the school facilities and student achievement.

Work Children's caregivers and, to a lesser extent, older siblings spend a significant percentage of their time in work environments outside of the child's direct experience. Extensive research over the last several decades has documented impacts of work environments on adult physical and mental health (Kuper & Marmot, 2003; Repetti et al., 2009). In addition, there is growing evidence for the impacts of work stress and other aspects of caregivers' work lives on their interactions with family members within their households, especially immediately after work (Repetti et al., 2009; Repetti & Wang, 2010). This stress, frequently stemming from work overload and/or negative interactions with supervisors and coworkers, might result in social withdrawal (and thus reduced social interactions and emotional responsiveness to family members), higher levels of family conflict, and lower levels of warmth and acceptance. These impacts have been demonstrated in parent-parent interactions and in caregiver-child interactions from preschool through adolescence. Also, more negative workplace environments for parents of infants predict higher levels of negative parenting behaviors and lower levels of positive parenting behaviors (Costigan et al., 2003).

Beyond workplace stress, the level of autonomy and cognitive complexity of parents' jobs is associated with parenting practices. Parents employed in positions with more autonomy and greater complexity demonstrate less restrictive and less direct parental control, less authoritarian and harsh parenting, and greater parental acceptance, warmth, and responsiveness (Greenberger et al., 1994; Grimm-Thomas & Perry-Jenkins, 1994; Mason et al., 1994; Parcel & Menaghan, 1993; Whitbeck et al., 1997). However, what is driving these associations is unclear in that parents with particular characteristics might both seek out and be well qualified for positions with greater autonomy and cognitive complexity.

Additional aspects of work environments that have demonstrated impacts on family interactions and processes include time at work, work hours in relation to children's school and childcare hours, predictability of work schedules, and job loss and unemployment (Repetti & Wang, 2010; Roeters et al., 2010). All of these factors have shown or have potential to demonstrate impacts on both the quantity and, more importantly, the quality of time families spend together. Unpredictable work hours, nonstandard work hours, time at work, and longer commutes impact family opportunities to engage in important routines and rituals, such as family mealtimes (see also Ferguson & Evans, 2019; Fiese, 2006; Fiese & Winter, 2010), and job loss and unemployment can impact family stability and overall functionality. Commuting stress also has direct impacts on adult mental health and thus likely impacts caregiver-child interactions post-work (Wener & Evans, 2011). Commuting can be particularly stressful for persons with childcare responsibilities because they are under greater time pressure to pick up children and complete household chores (Feng & Boyle, 2014; Wener & Evans, 2011). Further research to better understand the mechanisms underlying work impacts on family interactions and processes is clearly warranted (Repetti & Wang, 2010).

Cumulative Environmental Risks and Vulnerable Families

As has been documented in this chapter, the physical environment exerts both direct and indirect effects on children and their families over time, impacting multiple components of child health and well-being through direct impacts on the child and through impacts on family interactions and processes. It is rarely the case that an individual child or family is only exposed to one physical environment risk factor, for example, crowding, independent of other risk factors such as noise, chaos, and air pollution. Indeed, low-income children and families; BIPOC children; immigrant, refugee, unhoused, and otherwise displaced children; and children growing up in the Global South are disproportionately more likely to be exposed to multiple physical environment risk factors *at a given point in time*, as well as more likely to be exposed to multiple risk factors *over time* (Evans, 2006; Evans et al., 2012a, b; Evans et al., 2013; Evans & English, 2002; Ferguson et al., 2013; Ferguson & Evans, 2019). It is for this reason that a bioecocultural approach based on Bronfenbrenner's bioecological model (see Fig. 7.1) can help us to better understand the multiple intersecting and cumulative impacts of poor quality physical environments on children and their families over time.

Child and family exposure to multiple risk factors can have cumulative impacts on family interactions and processes through several pathways (Doan et al., 2012; Evans et al., 2013; Evans & English, 2002; Evans & Kim, 2013). Vulnerable children and families typically have higher exposure to cumulative risks across multiple physical settings while also having lower exposure to positive environments that might ameliorate some of these negative effects – a double jeopardy (Evans, 2004; Ferguson & Evans, 2019). Furthermore, all members of vulnerable households, including target children, their siblings, and their caregivers, likely have increased vulnerability to the impacts of negative physical environments as a result of prior, potentially chronic, undernutrition, compromised immune systems via hyper-inflammatory responses resulting from asthma and other chronic health conditions, and higher allostatic loads resulting from chronic stress. Taken together, multiple cumulative risk factors for individual family members within vulnerable families result in fewer positive caregiver-child and sibling-child interactions, a greater number of negative family interactions, and the disruption of effective proximal processes as a result of chaotic living situations (Evans et al., 2013; Ferguson & Evans, 2019).

It should also be noted here that the number and type of cumulative risks experienced by individual children and their families within and beyond their household environments might vary significantly across and within contexts. Thus, although the extant research in the Global South to date overall demonstrates similar impacts of individual components of the physical environment on child, adolescent, and adult health and well-being, the cumulative experiences of children and their families growing up in diverse global contexts might vary considerably. Thus, ongoing work to effectively study the multiple intersections of cumulative risks and protective factors across multiple global contexts is essential.

Limitations and Future Directions

In this chapter, we discuss some of the impacts of the physical environment (toxins and pollutants, food and water insecurity, lighting and temperature, noise, crowding, chaos, and housing quality) on the family interactions and processes of children and adolescents within home environments. In addition, we discuss these and other key physical environment influences on caregiver-child, sibling-child, and peer-child interactions in settings that are extensions of home, including the immediate neighborhood, play spaces, afterschool sites, and youth centers. Finally, we discuss two types of setting beyond home that are occupied by the child, siblings, and/or caregivers for a significant percentage of time, namely, child and sibling school and childcare settings and sibling and parent work settings. Where possible, we document the impacts of the physical environment on family interactions and processes, noting impacts on parenting practices, attachment, caregiver-child relationships and interactions, sibling relationships, family routines and rituals, family conflict, interpersonal aggression, child maltreatment, and social support (including social capital and social networks).

Our review has documented extensive impacts of the physical environment on child health and well-being, including physical and mental health, physiological stress, and socioemotional and cognitive development and functioning. In addition, there is growing evidence for the impacts of the physical environment on family interactions, and particularly on parenting practices and caregiver-child interactions. There is also an emerging literature investigating the impacts of the physical environment on key family processes, including routines and rituals such as family mealtimes. Nevertheless, the majority of this work to date has studied urban children and families living in the Global North. In addition, although a significant body of work to date has investigated the impacts of the physical environment on low-income children, less is known for BIPOC children, as well as immigrant, refugee, unhoused, and otherwise displaced children. Furthermore, much of the extant work looks at the impacts of individual contributing factors, yet we know that vulnerable children and families face multiple cumulative risk factors, and a smaller number of protective factors, within their home, neighborhood, school, afterschool, play, and work environments. Finally, research investigating resilience in the face of challenging physical environments tends to focus on personal rather than family and community factors without sufficient considerations of promotive environmental conditions that might protect or buffer children and families against otherwise negative physical environments (Masten, 2001). Where this work exists, there is limited consideration of large-scale, macro-events and processes that have the potential to significantly disrupt family interactions and processes on both a short- and long-term basis. These include immigration and forced migration, climate change and natural disasters, and health pandemics, many of which co-occur for vulnerable families.

A recent evaluation of the progress to date on the Sustainable Development Goals impacting early childhood makes the case for establishing interventions that are shown to be successful, smart, and sustainable (Britto et al., 2017). In other words, the implementation of multi-sectoral interventions that are focused on nurturing care, or a stable physical and social environment that meets children's health and nutrition needs, provides protections from threats and creates opportunities for positive interactions and learning for children, caregivers, and families.

It is essential to consider the impacts of multiple risk factors associated with family poverty, including stress and chaos, on effective proximal processes. Proximal processes are defined as regularly occurring, sustained reciprocal interactions between the developing child and the persons, objects, and symbols within their immediate physical and social context, including caregivers and siblings within home environments (Ferguson & Evans, 2019). Several potential intervention strategies for families are recommended, including Community Adventure Play Experiences (CAPEs; CDI, 2021; Ferguson et al., 2017) and community-based family literacy programs. Family literacy programs and CAPEs (temporary play spaces within children's own communities that engage them in interactive play with recycled and natural materials, ideally within natural settings) have the potential to increase positive social interactions between vulnerable children, peers, siblings, and their adult caregivers at low or no cost. When these types of programs are truly community-based and participatory, they have the potential of empowering all family members to actively combat the negative impacts of poor quality physical and social environments in their homes, neighborhoods, and schools through direct, sustainable, and low-cost community action. Furthermore, they provide an opportunity for the establishment of family routines and rituals around shared play, child-directed play, and shared book reading.

Beyond family- and community-level interventions, macro-level interventions that both support positive family interactions and the effective establishment of supportive family routines and rituals and reduce negative family interactions within home, neighborhood, and school environments are essential (Britto et al., 2017; Ferguson & Evans, 2019). Key policies that would make a significant difference for vulnerable families, and indeed all families, include universal high-quality childcare and afterschool care, a minimum living wage for teens and adults, flexible family leave policies, universal health insurance, mandates on the accessibility of affordable housing in both rural and urban communities, and significant broad-spectrum government-funded improvements in basic infrastructure, including sanitation, running water, and accessible, sustainable energy. Clearly, much work is still needed if we are to create healthy physical and social environments in which children and their families can thrive.

References

Aarts, M. J., de Vries, S. I., Van Oers, H. A., & Schuit, A. J. (2012). Outdoor play among children in relation to neighborhood characteristics: A cross-sectional neighborhood observation study. *International Journal of Behavioral Nutrition and Physical Activity, 9*(1), 1–11. https://doi.org/10.1186/1479-5868-9-98

Ackerman, B. P., & Brown, E. D. (2010). Physical and psychosocial turmoil in the home and cognitive development. In G. W. Evans & T. D. Wachs (Eds.), *Chaos and its influence on children's development* (pp. 35–48). American Psychological Association. https://doi.org/10.1037/12057-003

Akram, F., & Shamama-tus-Sabah, S. (2020). Home chaos: Predictor of behavioral and peer problems among young children. *Pakistan Journal of Psychological Research, 35*(3), 509–521. https://doi.org/10.33824/PJPR.2020.35.3.27

Alparone, F. R., & Pacilli, M. G. (2012). On children's independent mobility: The interplay of demographic, environmental, and psychosocial factors. *Children's Geographies, 10*(1), 109–122. https://doi.org/10.1080/14733285.2011.638173

Andrews, K., Atkinson, L., Harris, M., & Gonzalez, M. (2021). Examining the effects of household chaos on child executive functions: A meta-analysis. *Psychological Bulletin, 147*(1), 16–32. https://doi.org/10.1037/bul0000311

Appleyard, D., & Lintell, M. (1972). The environmental quality of city streets: The residents' viewpoint. *Journal of the American Institute of Planners, 38*(2), 84–101. https://doi.org/10.1080/01944367208977410

Aschengrau, A., Weinberg, J. M., Janulewicz, P. A., Romano, M. E., Gallagher, L. G., Winter, M. R., Martin, B. R., Vieira, V. M., Webster, T. F., White, R. F., & Ozonoff, D. M. (2011). Affinity for risky behaviors following prenatal and early childhood exposure to tetrachloroethylene (PCE)-contaminated drinking water: A retrospective cohort study. *Environmental health: A Global access science source, 10*(1), 102. https://doi.org/10.1186/1476-069X-10-102

Banna, H. U., Anjum, A., Biswas, S., Mondal, V., Siddique, A. E., Roy, A. K., Nikkon, F., Haque, A., Himeno, S., Salam, K. A., Hossain, K., & Saud, Z. A. (2022). Parental lead exposure promotes neurobehavioral disorders and hepatic dysfunction in mouse offspring. *Biological Trace Element Research, 200*(3), 1171–1180. https://doi.org/10.1007/s12011-021-02709-y

Bannerjee, K., & Driskell, D. (2002). India tales from truth town. In L. Chawla (Ed.), *Growing up in an urbanising world* (pp. 135–160). Earthscan/UNESCO.

Barbone, F., Rosolen, V., Mariuz, M., Parpinel, M., Casetta, A., Sammartano, F., Ronfani, L., Brumatti, L. V., Bin, M., Castriotta, L., Valent, F., Little, D. L., Mazej, D., Tratnik, J. S., Višnjevec, A. M., Sofianou, K., Špirić, Z., Krsnik, M., Osredkar, J., et al. (2019). Prenatal mercury exposure and child neurodevelopment outcomes at 18 months: Results from the Mediterranean PHIME cohort. *International Journal of Hygiene and Environmental Health, 222*(1), 9–21. https://doi.org/10.1016/j.ijheh.2018.07.011

Bartlett, S. (1999). Children's experience of the physical environment in poor urban settlements and the implications for policy, planning and practice. *Environment & Urbanization, 11*(2), 63–73. https://doi.org/10.1177/2F095624789901100207

Bartlett, S. (2003). Water, sanitation and urban children: The need to go beyond "improved" provision. *Environment & Urbanization, 15*(2), 57–70. https://doi.org/10.1177/095624780301500220

Bartlett, S. (2008). The implications of climate change for children in lower-income countries. *Children, Youth & Environments, 18*(1), 72–98. https://www.researchgate.net/publication/284682338_The_implications_of_climate_change_for_children_in_lower-income_countries

Bartlett, S., Hart, R., Satterthwaite, D., de la Barra, X., & Missair, A. (1999). Cities for children: Children's rights, poverty and urban management. *Earthscan/UNESCO*.

Beery, A. K., & Kaufer, D. (2015). Stress, social behavior, and resilience: Insights from rodents. *Neurobiology of Stress, 1*, 116–127. https://doi.org/10.1016/j.ynstr.2014.10.004

Bell, A. C., & Dyment, J. E. (2008). Grounds for health: The intersection of green school grounds and health-promoting schools. *Environmental Education Research, 14*, 77–90. https://doi.org/10.1080/13504620701843426

Bernal, G., Morales-Cruz, J., & Gómez-Arroyo, K. (2019). In search of a global family psychology: Challenges and perspectives. In B. H. Fiese, M. Celano, K. Deater-Deckard, E. N. Jouriles, & M. A. Whisman (Eds.), *APA handbook of contemporary family psychology: Applications and broad impact of family psychology* (pp. 605–623). American Psychological Association. https://doi.org/10.1037/0000100-037

Booker, J. M., & Hellekson, C. J. (1992). Prevalence of seasonal affective disorder in Alaska. *American Journal of Psychiatry, 149*(9), 1176–1182. https://doi.org/10.1176/ajp.149.9.1176

Booth, A., & Edwards, J. N. (1976). Crowding and family relations. *American Sociological Review, 41*(2), 308–321. https://doi.org/10.2307/2094476

Bradley, R. H., & Caldwell, B. M. (1984). The HOME inventory and family demographics. *Developmental Psychology, 20*(2), 315–320. https://doi.org/10.1037/0012-1649.20.2.315

Bradley, R. H., Caldwell, B. M., & Corwyn, R. F. (2003). The child care HOME inventories: Assessing the quality of family child care homes. *Early Childhood Research Quarterly, 18*(3), 294–309. https://doi.org/10.1016/S0885-2006(03)00041-3

Britto, P. R., Lye, S. J., Proulx, K., Yousafzai, A. K., Matthews, S. G., Vaivada, T., Perez-Escamilla, R., Rao, N., Ip, P., Fernald, L. C. H., MacMillan, H., Hanson, M., Wachs, T. D., Yao, H., Yoshikawa, H., Cerezo, A., & The Early Childhood Development Interventions Review Group. (2017). Nurturing care: Promoting early childhood development. *The Lancet, 389*(10064), 91–102. https://doi.org/10.1016/S0140-6736(16)31390-3

Bronfenbrenner, U. (1977). Toward an experimental ecology of human development. *American Psychologist, 32*(7), 513–531. https://doi.org/10.1037/0003-066X.32.7.513

Bronfenbrenner, U. (1986). Ecology of the family as a context for human development: Research perspectives. *Developmental Psychology, 22*(6), 723–742. https://doi.org/10.1037/0012-1649.22.6.723

Bronfenbrenner, U., & Ceci, S. J. (1994). Nature-nurture reconceptualized in developmental perspective: A bioecological model. *Psychological Review, 101*(4), 568–586. https://doi.org/10.1037/0033-295x.101.4.568

Bronfenbrenner, U., & Crouter, A. C. (1983). The evolution of environmental models in developmental research. In W. Kessen & P. H. Mussen (Eds.), *Handbook of child psychology: Vol. 1. History, theory, and methods* (4th ed., pp. 357–314). Wiley.

Bronfenbrenner, U., & Evans, G. W. (2000). Developmental science in the 21st century: Emerging theoretical models, research designs, and empirical findings. *Social Development, 9*, 115–125. https://doi.org/10.1111/1467-9507.00114

Bronfenbrenner, U., & Morris, P. A. (2006). The bioecological model of human development. In R. M. Lerner (Ed.), *Handbook of child development* (Vol. 1. *Theoretical models of human development*, 6th ed., pp. 793–828). Wiley.

Browning, C. R., Bjornstrom, E. E., & Cagney, K. A. (2011). Health and mortality consequences of the physical environment. In R. G. Rogers & E. M. Crimmins (Eds.), *International handbook of adult mortality* (pp. 441–464). Springer.

Bruce, N. G., Dherani, M. K., Das, J. K., Balakrishnan, K., Adair-Rohani, H., Bhutta, Z. A., & Pope, D. (2013). Control of household air pollution for child survival: Estimates for intervention impacts. *BMC Public Health, 13*(Suppl 3), 1–13. https://doi.org/10.1186/2F1471-2458-13-S3-S8

Buckley, J., Schneider, M., & Shang, Y. (2004). *The effects of school facility quality on teacher retention in urban school districts*. https://files.eric.ed.gov/fulltext/ED539484.pdf

Bures, R. M. (2003). Childhood residential stability and health at midlife. *American Journal of Public Health, 93*(7), 1144–1148. https://doi.org/10.2105/ajph.93.7.1144

Calhoun, J. B. (1962). Population density and social pathology. *Scientific American, 206*, 139–148. https://www.gwern.net/docs/sociology/1962-calhoun.pdf

Carlson, J. A., Saelens, B. E., Kerr, J., Schipperijn, J., Conway, T. L., Frank, L. D., Chapman, J. E., Glanz, K., Cain, K. L., & Sallis, J. F. (2015). Association between neighborhood walkability and GPS-measured walking, bicycling and vehicle time in adolescents. *Health & Place, 32*, 1–7. https://doi.org/10.1016/2Fj.healthplace.2014.12.008

Carter-Pokras, O., Zambrana, R. E., Poppell, C. F., Logie, L. A., & Guerrero-Preston, R. (2007). The environmental health of Latino children. *Journal of Pediatric Health Care, 21*(5), 307–314. https://doi.org/10.1016/2Fj.pedhc.2006.12.005

Cash, C. S. (1993). *Building condition and student achievement and behavior* [doctoral dissertation, Virginia Polytechnic Institute and State University]. Building condition and student achievement and behavior (https://vtechworks.lib.vt.edu/handle/10919/29200).

Chatterjee, S. (2007). Children's role in humanizing forced evictions and resettlements in Delhi. *Children, Youth and Environments, 17*(1), 198–221.

Chawla, L. (2002). Cities for human development. In L. Chawla (Ed.), *Growing up in an urbanizing world* (pp. 15–34). Earthscan/UNESCO.

Chawla, L. (2015). Benefits of nature contact for children. *Journal of Planning Literature, 30*(4), 433–452. https://doi.org/10.1177/2F0885412215595441

Child Development Institute (CDI). (2021). *Community Adventure Play Experience (CAPE)*. [Community outreach] Sarah Lawrence College. https://www.sarahlawrence.edu/cdi/outreach/

Chin, S. Y., Hopkins, W. A., & Cristol, D. A. (2017). Mercury alters initiation and construction of nests by zebra finches, but not incubation or provisioning behaviors. *Ecotoxicology, 26*(9), 1271–1283. https://doi.org/10.1007/s10646-017-1852-x

Chithra, V. S., & Madanayak, S. N. S. (2018). Source identification of indoor particulate matter and health risk assessment in school children. *Journal of Hazardous, Toxic, and Radioactive Waste, 22*(2), 04018002. https://doi.org/10.1061/(ASCE)HZ.2153-5515.0000390

Christian, J. J., & LeMunyan, C. D. (1958). Adverse effects of crowding on lactation and reproduction of mice and two generations of their progeny. *Endocrinology, 63*(5), 517–529. https://doi.org/10.1210/endo-63-5-517

Christian, H., Zubrick, S. R., Foster, S., Giles-Corti, B., Bull, F., Wood, L., Knuiman, M., Brinkman, S., Houghton, S., & Boruff, B. (2015). The influence of the neighborhood physical environment on early child health and development: A review and call for research. *Health & Place, 33*, 25–36. https://doi.org/10.1016/j.healthplace.2015.01.005

Clayton, S. D. (2012). Environment and identity. In S. D. Clayton (Ed.), *The Oxford handbook of environmental and conservation psychology* (pp. 164–180). Oxford University Press. https://doi.org/10.1093/oxfordhb/9780199733026.013.0010

Cohen, S., Evans, G. W., Stokols, D., & Krantz, D. S. (1986). Stress processes and the costs of coping. In *Behavior, health, and environmental stress* (pp. 1–23). Springer. https://doi.org/10.1007/978-1-4757-9380-2_1

Cohen, D. A., Inagami, S., & Finch, B. (2008). The built environment and collective efficacy. *Health & Place, 14*(2), 198–208. https://doi.org/10.1016/j.healthplace.2007.06.001

Cook, J. T., Frank, D. A., Berkowitz, C., Black, M. M., Casey, P. H., Cutts, D. B., Meyers, A. F., Zaldivar, N., Skalicky, A., Levenson, S., Heeren, T., & Nord, M. (2004). Food insecurity is associated with adverse health outcomes among human infants and toddlers. *The Journal of Nutrition, 134*(6), 1432–1438. https://doi.org/10.1093/jn/134.6.##

Coplan, J. D., Gupta, N. K., Karim, A., Rozenboym, A., Smith, E. L. P., Kral, J. G., & Rosenblum, A. L. (2017). Maternal hypothalamic-pituitary-adrenal axis response to foraging uncertainty: A model of individual *vs.* social allostasis and the "superorganism hypothesis". *PLoS One, 12*(9), e0184340. https://doi.org/10.1371/journal.pone.0184340

Corapci, F., & Wachs, T. D. (2002). Does parental mood or efficacy mediate the influence of environmental chaos upon parenting behavior? *Merrill-Palmer Quarterly, 48*(2), 182–201. https://doi.org/10.1353/mpq.2002.0006

Cosco, N., & Moore, R. C. (2002). Our neighbourhood is like that! Cultural richness and childhood identity in Boca Baraccas, Buenos Aries. In L. Chawla (Ed.), *Growing up in an urbanizing world* (pp. 35–56). Earthscan/UNESCO.

Costigan, C. L., Cox, M. J., & Cauce, A. M. (2003). Work-parenting linkages among dual-earner couples at the transition to parenthood. *Journal of Family Psychology, 17*(3), 397–408. https://doi.org/10.1037/0893-3200.17.3.397

Dalton, P., Claeson, A. S., & Horenziak, S. (2020). The impact of indoor malodor: Historical perspective, modern challenges, negative effects, and approaches for mitigation. *Atmosphere, 11*(2), 126. https://doi.org/10.3390/atmos11020126

de Dear, R., Kim, J., Candido, C., & Deuble, M. (2015). Adaptive thermal comfort in Australian school classrooms. *Building Research & Information, 43*(3), 383–398. https://doi.org/10.1080/09613218.2015.991627

Diez Roux, A. V., & Mair, C. (2010). Neighborhoods and health. *Annals of the New York Academy of Sciences, 1186*, 125–145. https://doi.org/10.1111/j.1749-6632.2009.05333.x

Dizon, A. M., & Quijano, S. (1997). Impact of eviction on children. *Report for Urban Poor Associates, Asian Coalition for Housing Rights (ACHR) and United Nations Economic and Social Commission for Asia and the Pacific (UN-ESCA).*

Doan, S. N., & Evans, G. W. (2020). Chaos and instability from birth to age three. *Future of Children, 30*(2), 93–113. EJ1293542.pdf (ed.gov)

Doan, S. N., Fuller-Rowell, T. E., & Evans, G. W. (2012). Cumulative risk and adolescent's internalizing and externalizing problems: The mediating roles of maternal responsiveness and self-regulation. *Developmental Psychology, 48*(6), 1529. https://doi.org/10.1037/a0027815

Earthman, G. I. (2004). Prioritization of 31 criteria for school building adequacy. *American Civil Liberties Union Foundation of Maryland.*. https://www.researchgate.net/publication/239605533_Prioritization_of_31_criteria_for_school_building_adequacy

Edwards, B., & Bromfield, L. M. (2009). Neighborhood influences on young children's conduct problems and pro-social behavior: Evidence from an Australian national sample. *Children and Youth Services Review, 31*(3), 317–324. https://doi.org/10.1016/j.childyouth.2008.08.005

Edwards, J. N., Fuller, T. D., Sermsri, S., & Vorakitphokatorn, S. (1994). Why people feel crowded: An examination of objective and subjective crowding. *Population and Environment, 16*(2), 149–173. https://doi.org/10.1007/BF02208781

Ehrenberg, R. G., Brewer, D. J., Gamoran, A., & Willms, J. D. (2001). Class size and student achievement. *Psychological Science in the Public Interest, 2*(1), 1–30. https://doi.org/10.1111/2F1529-1006.003

Evans, G. W. (2003). The built environment and mental health. *Journal of Urban Health, 80*(4), 536–555. https://doi.org/10.1093/jurban/jtg063

Evans, G. W. (2004). The environment of childhood poverty. *American Psychologist, 59*(2), 77–92. https://doi.org/10.1037/0003-066X.59.2.77

Evans, G. W. (2006). Child development and the physical environment. *Annual Review of Psychology, 57*, 423–451. https://doi.org/10.1146/annurev.psych.57.102904.190057

Evans, G. W. (2021). The physical context of child development. *Current Directions in Psychological Science, 30*(1), 41–48. https://doi.org/10.1177/2F0963721420980719

Evans, G. W., & English, K. (2002). The environment of poverty: Multiple stressor exposure, psychophysiological stress, and socioemotional adjustment. *Child Development, 73*(4), 1238–1248. https://doi.org/10.1111/1467-8624.00469

Evans, G. W., Fuller-Rowell, T. E., & Doan, S. N. (2012a). Childhood cumulative risk and obesity: The mediating role of self-regulatory ability. *Pediatrics, 129*(1), e68–e73. https://doi.org/10.1542/peds.2010-3647

Evans, G. W., Jones-Rounds, M. L., Belojevic, G., & Vermeylen, F. (2012b). Family income and childhood obesity in eight European cities: The mediating roles of neighborhood characteristics and physical activity. *Social Science & Medicine, 75*(3), 477–481. https://doi.org/10.1016/j.socscimed.2012.03.037

Evans, G. W., & Kim, P. (2013). Childhood poverty, chronic stress, self-regulation and coping. *Child Development Perspectives, 7*, 43–48. https://doi.org/10.1111/cdep.12013

Evans, G. W., & Lepore, S. J. (1993). Nonauditory effects of noise on children: A critical review. *Children's Environments, 10*(1), 31–51. http://www.jstor.org/stable/41515250

Evans, G. W., Lepore, S. J., Shejwal, B. R., & Palsane, M. N. (1998). Chronic residential crowding and children's Well-being: An ecological perspective. *Child Development, 69*(6), 1514–1523. https://doi.org/10.2307/1132129

Evans, G. W., Li, D., & Whipple, S. S. (2013). Cumulative risk and child development. *Psychological Bulletin, 139*(6), 1342–1396. https://doi.org/10.1037/a0031808

Evans, G. W., Maxwell, L. E., & Hart, B. (1999). Parental language and verbal responsiveness to children in crowded homes. *Developmental Psychology, 35*(4), 1020–1023. https://doi.org/10.1037/0012-1649.35.4.1020

Evans, G. W., Ricciuti, H. N., Hope, S., Schoon, I., Bradley, R. H., Corwyn, R. F., & Hazan, C. (2010). Crowding and cognitive development: The mediating role of maternal responsiveness among 36-month-old children. *Environment and Behavior, 42*(1), 135–148. https://doi.org/10.1177/2F0013916509333509

Evans, G. W., Saltzman, H., & Cooperman, J. L. (2001). Housing quality and children's socioemotional health. *Environment and Behavior, 33*(3), 389–399. https://doi.org/10.1177/2F00139160121973043

Evans, G. W., & Stecker, R. (2004). Motivational consequences of environmental stress. *Journal of Environmental Psychology, 24*(2), 143–165. https://doi.org/10.1016/S0272-4944(03)00076-8

Evans, G. W., & Wachs, T. D. (Eds.). (2010). *Chaos and its influence on children's development: An ecological perspective*. American Psychological Association.

Evans, G. W., Wells, N. M., & Moch, A. (2003). Housing and mental health: A review of the evidence and a methodological and conceptual critique. *Journal of Social Issues, 59*(3), 475–500. https://doi.org/10.1111/1540-4560.00074

Faber Taylor, A., Wiley, A., Kuo, F. E., & Sullivan, W. C. (1998). Growing up in the inner city: Green spaces as places to grow.. *Environment and Behavior*, *30*(1), 3-27. 10.1177%2F0013916598301001

Faber Taylor, A., Kuo, F. E., & Sullivan, W. C. (2002). Views of nature and self-discipline: Evidence from inner city children. *Journal of Environmental Psychology, 22*(1–2), 49–63. https://doi.org/10.1006/jevp.2001.0241

Fan, Y., & Chen, Q. (2012). Family functioning as a mediator between neighborhood conditions and children's health: Evidence from a national survey in the United States. *Social Science & Medicine, 74*(12), 1939–1947. https://doi.org/10.1016/j.socscimed.2012.01.039

Feng, Z., & Boyle, P. (2014). Do long journeys to work have adverse effects on mental health. *Environment and Behavior, 46*(5), 609–625. https://doi.org/10.1177/2F0013916512472053

Ferguson, K. T. (2008). *Infant development in the context of Malawian infant homes*. [Unpublished doctoral dissertation.]. Cornell University.

Ferguson, K. T., Cassells, R. C., MacAllister, J. W., & Evans, G. W. (2013). The physical environment and child development: An international review. *International Journal of Psychology, 48*(4), 437–468. https://doi.org/10.1080/00207594.2013.804190

Ferguson, K. T., & Evans, G. W. (2019). Family systems and family psychology in bioecological and bioecocultural perspective. In B. H. Fiese (Ed.), *APA handbook of contemporary family psychology* (Vol. I, pp. 143–161). American Psychological Association.

Ferguson, K. T., & Lee, M. J. (2013). Cognitive, motor and behavioral development of orphans of HIV/AIDS in institutional contexts. In M. Boivin & B. Giordani (Eds.), *Neuropsychology of children in Africa: Risk and resilience within a co-constructivist paradigm* (pp. 69–93). Springer. https://doi.org/10.1007/978-1-4614-6834-9_4

Ferguson, K. T., Schecter, B., Feltham, E., & Davis, A. (2017). Play's the thing: Community adventure play experiences. *IPA/USA E-Journal*, 15–20. https://digitalcommons.slc.edu/cdi_research/1/

Fiese, B. H. (2006). Family routines and rituals. *Yale University Press*. https://yalebooks.yale.edu/book/9780300116960/family-routines-and-rituals

Fiese, B. H., Foley, K. P., & Spagnola, M. (2006). Routine and ritual elements in family mealtimes: Contexts for child Well-being and family identity. *New Directions for Child and Adolescent Development, 111*, 67–89. https://doi.org/10.1002/cd.156

Fiese, B. H., Jones, B. L., & Jarick, J. M. (2015). Family mealtime dynamics and food consumption: An experimental approach to understanding distractions. *Couple and Family Psychology: Research and Practice, 4*(4), 199–211. https://doi.org/10.1037/cfp0000047

Fiese, B. H., Jones, B. L., & Saltzman, J. A. (2019). Systems unify family psychology. In B. H. F. M. Celano, K. Deater-Deckard, E. N. Jouriles, & M. A. Whisman (Eds.), *APA handbook of contemporary family psychology* (Vol. I, pp. 3–19). *American Psychological Association*. https://doi.org/10.1037/0000099-001

Fiese, B. H., & Schwartz, M. (2008). Reclaiming the family table: Mealtimes and child health and wellbeing. *Social Policy Report, 42*(4), 1–20. https://doi.org/10.1002/j.2379-3988.2008.tb00057.x

Fiese, B. H., & Winter, M. A. (2010). The dynamics of family chaos and its relation to children's socioemotional Well-being. In G. W. Evans & T. D. Wachs (Eds.), *Chaos and its influence on children's development: An ecological perspective* (pp. 49–66). *American Psychological Association*. https://doi.org/10.1037/12057-004

Fjørtoft, I. (2004). Landscape as playscape: The effects of natural environments on children's play and motor development. *Children, Youth and Environments, 14*(2), 21–44. https://www.researchgate.net/publication/252182057_Landscape_as_Playscape_The_Effects_of_Natural_Environments_on_Children%27s_Play_and_Motor_Development

Furlong, M. A., Engel, S. M., Barr, D. B., & Wolff, M. S. (2014). Prenatal exposure to organophosphate pesticides and reciprocal social behavior in childhood. *Environment International, 70*, 125–131. https://doi.org/10.1016/j.envint.2014.05.011

Gauvain, M., & Munroe, R. L. (2009). Contributions of societal modernity to cognitive development: A comparison of four cultures. *Child Development, 80*(6), 1628–1642. https://doi.org/10.1111/j.1467-8624.2009.01358.x

Gibson, J. J. (1979). The ecological approach to visual perception. *Houghton Mifflin*. https://doi.org/10.4324/9781315740218

Gifford, R. (2014). Environmental psychology matters. *Annual Review of Psychology, 65*, 541–579. https://doi.org/10.1146/annurev-psych-010213-115048

Giles-Corti, B., Kelty, S. F., Zubrick, S. R., & Villanueva, K. P. (2009). Encouraging walking for transport and physical activity in children and adolescents: How important is the built environment? *Sports Medicine, 39*(12), 995–1009. https://doi.org/10.2165/11319620-000000000-00000

Gill, T. (2014). The benefits of children's engagement with nature: A systematic literature review. *Children, Youth and Environments, 24*(2), 10–34. https://doi.org/10.7721/chilyoutenvi.24.2.0010

Glewwe, P., Hanushek, E. A., Humpage, S., & Ravina, R. (2013). School resources and educational outcomes in developing countries: A review of the literature from 1990 to 2010. In P. Glewwe (Ed.), *Education policy in developing countries* (pp. 13–64). University of Chicago Press.

Grantham-McGregor, S. M., Chang, S., & Walker, S. P. (1998). Evaluation of school feeding programs: Some Jamaican examples. *The American Journal of Clinical Nutrition, 67*(4), 785S–789S. https://doi.org/10.1093/ajcn/67.4.785s

Grantham-McGregor, S., Cheung, Y. B., Cueto, S., Glewwe, P., Richter, L., Strupp, B., & The International Child Development Steering Group. (2007). Developmental potential in the first 5 years for children in developing countries. *Lancet, 369*(9555), 60–70. https://doi.org/10.1016/S0140-6736(07)60032-4

Greenberger, E., O'Neil, R., & Nagel, S. K. (1994). Linking workplace and homeplace: Relations between the nature of adults' work and their parenting behaviors. *Developmental Psychology, 30*(6), 990–1002. https://doi.org/10.1037/0012-1649.30.6.990

Greenwald, R., Hedges, L. V., & Laine, R. D. (1996). The effect of school resources on student achievement. *Review of Educational Research, 66*(3), 361–396. https://doi.org/10.2307/1170528

Grimm-Thomas, K., & Perry-Jenkins, M. (1994). All in a day's work: Job experiences, self-esteem, and fathering in working-class families. *Family Relations, 43*(2), 174–181. https://doi.org/10.2307/585320

Hamadani, J. D., Grantham-McGregor, S. M., Tofail, F., Nermell, B., Fängström, B., Huda, S. N., Yesmin, S., Raman, M., Vera-Hernández, M., Arifeen, S. E., & Vahter, M. (2010). Pre- and postnatal arsenic exposure and child development at 18 months of age: A cohort study in rural Bangladesh. *International Journal of Epidemiology, 39*(5), 1206–1216. https://doi.org/10.1093/ije/dyp369

Hardoy, J. E., Mitlin, D., & Satterthwaite, D. (2001). *Environmental problems in an urbanizing world*. Routledge.

Harvey, P. W., & Chevins, P. F. D. (1987). Crowding during pregnancy delays puberty and alters estrous cycles of female offspring in mice. *Experientia, 43*(3), 306–308. https://doi.org/10.1007/bf01945561

Hattie, J., Marsh, H. W., Neill, J., & Richards, G. (1997). Adventure education and outward bound: Out of class experiences that make a lasting difference. *Review of Educational Research, 67*, 43–87. https://doi.org/10.3102/2F00346543067001043

Heft, H. (1979). Background and focal environmental conditions of the home and attention in young children. *Journal of Applied Social Psychology, 9*(1), 47–69. https://doi.org/10.1111/j.1559-1816.1979.tb00794.x

Heft, H. (1988). Affordances of children's environments: A functional approach to environmental description. *Children's Environments Quarterly, 5*(3), 29–37. https://doi.org/10.1006/jevp.2001.0249

Heschong, L. (1999). *Daylighting in schools: An investigation into the relationship between daylighting and human performance*. [report] submitted by the Heschong Mahone group to Pacific Gas and Electric, on behalf of the California Board for Energy Efficiency Third Party Program. https://doi.org/10.13140/RG.2.2.31498.31683

Higgins, S., Hall, E., Wall, K., Woolner, P., & McCaughey, C. (2005). *The impact of school environments: A literature review*. School of Education, Communication and Language Science, University of Newcastle.

Hiroto, D. S. (1974). Locus of control and learned helplessness. *Journal of Experimental Psychology, 102*(2), 187–193. https://doi.org/10.1037/h0035910

Homel, R., & Burns, A. (1989). Environmental quality and the well-being of children. *Social Indicators Research, 21*(2), 133–158. https://doi.org/10.1007/BF00300500

Hoy, W. K., & Sabo, D. J. (1997). *Quality middle schools: Open and healthy*. Corwin Press. Quality Middle Schools | SAGE Publications Inc.

Hüttenmoser, M. (1995). Children and their living surroundings: Empirical investigations into the significance of living surroundings for the everyday life and development of children. *Children's Environments, 12*(4), 403–413. https://www.jstor.org/stable/41514991

Hyland, C., Bradshaw, P., Deardorff, J., Gunier, R. B., Mora, A. M., Kogut, K., Sagiv, S. K., Bradman, A., & Eskenazi, B. (2022). Interactions of agricultural pesticide use near home during pregnancy and adverse childhood experiences on adolescent neurobehavioral development in the CHAMACOS study. *Environmental Research, 204*(Pt A), 111908. https://doi.org/10.1016/j.envres.2021.111908

Irwin, L. G., Siddiqi, A., & Hertzman, G. (2007). Early child development: A powerful equalizer. Submitted to the World Health Organization's Commission on the Social Determinants of Health http://www.who.int/social_determinants/themes/earlychilddevelopment/en/index.html

Jackson, C., Vynnycky, E., Hawker, J., et al. (2013). School closures and influenza: Systematic review of epidemiological studies. *BMJ Open, 3*(2), e002149. https://doi.org/10.1136/bmjopen-2012-002149

Jelleyman, T., & Spencer, N. (2008). Residential mobility in childhood and health outcomes: A systematic review. *Journal of Epidemiology and Community Health, 62*(7), 584–592. https://doi.org/10.1136/jech.2007.060103

Jo, Y. I., Lee, J. L., & Koo, J. H. (2018). Effect of physical environment and programs on the social interaction of youth space users in Seoul in the case of pilot projects. *Sustainability, 10*(12), 4515. https://doi.org/10.3390/su10124515

Jones, P. C., Pendergast, L. L., Schaefer, B. A., Rasheed, M., Svensen, E., Scharf, R., Shrestha, R., Maphula, A., Roshan, R., Rasmussen, Z., Seidman, J. C., Murray-Kolb, L. E., & MAL-ED Network Investigators. (2017). Measuring home environments across cultures: Invariance of the HOME scale across eight international sites from the MAL-ED study. *Journal of School Psychology, 64*, 109–127. https://doi.org/10.1016/j.jsp.2017.06.001

Kaplan, R., & Kaplan, S. (1989). *The experience of nature*. Cambridge University Press.

Kaplan, S., & Talbot, J. F. (1983). Psychological benefits of a wilderness experience. In I. Altman & J. F. Wohlwill (Eds.), *Behavior and the natural environment* (pp. 163–203). Springer.

Kelz, C., Evans, G. W., & Röderer, K. (2015). The restorative effects of redesigning the schoolyard: A multi-methodological, quasi-experimental study in rural Austrian middle schools. *Environment and Behavior, 47*(2), 119–139. https://doi.org/10.1177/0013916513510528

Kirkorian, H. L., Pempek, T. A., Murphy, L. A., Schmidt, M. E., & Anderson, D. R. (2009). The impact of background television on parent–child interaction. *Child Development, 80*(5), 1350–1359. https://doi.org/10.1111/j.1467-8624.2009.01337.x

Klatte, M., Spilski, J., Mayerl, J., Möhler, U., Lachmann, T., & Bergström, K. (2017). Effects of aircraft noise on reading and quality of life in primary school children in Germany: Results from the NORAH study. *Environment and Behavior, 49*(4), 390–424. https://doi.org/10.1177/2F0013916516642580

Knibbs, L. D., de Waterman, A. M. C., Toelle, B. G., Guo, Y., Denison, L., Jalaludin, B., Marks, G. B., & Williams, G. M. (2018). The Australian child health and air pollution study (ACHAPS): A national population-based cross-sectional study of long-term exposure to outdoor air pollution, asthma, and lung function. *Environment International, 120*, 394–403. https://doi.org/10.1016/j.envint.2018.08.025

Krantz, P. J. (1974). *Ecological arrangements in the classroom. [Unpublished doctoral dissertation]*. University of Kansas.

Kruger, J. S. (2002). Children in a South African squatter camp gain and lose a voice. In L. Chawla (Ed.), *Growing up in an urbanizing world* (pp. 111–133). Earthscan/UNESCO.

Kruger, J. S., & Chawla, L. (2002). "We know something someone doesn't know": Children speak out on local conditions in Johannesburg. *Environment and Urbanization, 14*(2), 85–96. https://doi.org/10.1177/095624780201400207

Küller, R., & Lindsten, C. (1992). Health and behavior of children in classrooms with and without windows. *Journal of Environmental Psychology, 12*(4), 305–317. https://doi.org/10.1016/S0272-4944(05)80079-9

Kuo, F. E., & Faber Taylor, A. (2004). A potential natural treatment for attention-deficit/hyperactivity disorder: Evidence from a national study. *American Journal of Public Health, 94*(9), 1580–1586. https://doi.org/10.2105/AJPH.94.9.1580

Kuper, H., & Marmot, M. (2003). Job strain, job demands, decision latitude, and risk of coronary heart disease within the Whitehall II study. *Journal of Epidemiology & Community Health, 57*(2), 147–153. https://doi.org/10.1136/2Fjech.57.2.147

Kyttä, M. (2002). Affordances of children's environments in the context of cities, small towns, suburbs and rural villages in Finland and Belarus. *Journal of Environmental Psychology, 22*(1–2), 109–123. https://doi.org/10.1006/jevp.2001.0249

Kyttä, M. (2004). The extent of children's independent mobility and the number of actualized affordances as criteria for child-friendly environments. *Journal of Environmental Psychology, 24*(2), 179–198. https://doi.org/10.1016/S0272-4944(03)00073-2

Kyttä, M., Oliver, M., Ikeda, E., Ahmadi, E., Omiya, I., & Laatikainen, T. (2018). Children as urbanites: Mapping the affordances and behavior settings of urban environments for Finnish and Japanese children. *Children's Geographies, 16*(3), 319–332. https://doi.org/10.1080/14733285.2018.1453923

Landrigan, P. J., Fuller, R., Fisher, S., Suk, W. A., Sly, P., Chiles, T. C., & Bose-O'Reilly, S. (2019). Pollution and children's health. *Science of the Total Environment, 650*, 2389–2394. https://doi.org/10.1016/j.scitotenv.2018.09.375

Laughlin, N. K. (1986). Animal models of behavioral effects of early lead exposure. In E. P. Riley & C. V. Vorhess (Eds.), *Handbook of Behavioral teratology* (pp. 291–319). Plenum9. Handbook of Behavioral Teratology | SpringerLink.

Leventhal, T., & Brooks-Gunn, J. (2000). The neighborhoods they live in: The effects of neighborhood residence on child and adolescent outcomes. *Psychological Bulletin, 126*(2), 309–337. https://doi.org/10.1037/0033-2909.126.2.309

Liddell, C., & Kruger, P. (1987). Activity and social behavior in a South African township nursery: Some effects of crowding. *Merrill-Palmer Quarterly, 33*(2), 195–211.

Liddell, C., & Kruger, P. (1989). Activity and social behavior in a crowded South African township nursery: A follow-up study on the effects of crowding at home. *Merrill-Palmer Quarterly, 35*, 209–226. https://www.jstor.org/stable/23086329

Loebach, J., & Cox, A. (2020). Tool for observing play outdoors (TOPO): A new typology for capturing children's play behaviors in outdoor environments. *International Journal of Environmental Research and Public Health, 17*(15), 5611. https://doi.org/10.3390/2Fijerph17155611

Loebach, J., Little, S., Cox, A., & Owens, P. E. (Eds.). (2020). *The Routledge handbook of designing public spaces for young people: Processes, practices and policies for youth inclusion.* Routledge.

Loebach, J., Sanches, M., Jaffe, J., & Elton-Marshall, T. (2021). Paving the way for outdoor play: Examining the socio-emotional environmental barriers to community-based outdoor play. *International Journal of Environmental Research and Public Health, 18*(7), 3617. https://doi.org/10.3390/ijerph18073617

Macpherson, A., Roberts, I., & Pless, I. B. (1998). Children's exposure to traffic and pedestrian injuries. *American Journal of Public Health, 88*(12), 1840–1843. https://doi.org/10.2105/ajph.88.12.1840

Marcal, K. E. (2021). Pathways to adolescent emotional and behavioral problems: An examination of maternal depression and harsh parenting. *Child Abuse & Neglect, 113*, 104917. https://doi.org/10.1016/j.chiabu.2020.104917

Marsh, S., Dobson, R., & Maddison, R. (2020). The relationship between household chaos and child, parent, and family outcomes: A systematic scoping review. *BMC Public Health, 20*, 513. https://doi.org/10.1186/s12889-020-08587-8

Mason, C. A., Cauce, A. M., Gonzales, N., Hiraga, Y., & Grove, K. (1994). An ecological model of externalizing behaviors in African-American adolescents: No family is an island. *Journal of Research on Adolescence, 4*(4), 639–655. https://doi.org/10.1207/s15327795jra0404_12

Masten, A. S. (2001). Ordinary magic: Resilience processes in development. *American Psychologist, 56*(3), 227–238. https://doi.org/10.1037//0003-066x.56.3.227

Matheny, A. P., Jr., Wachs, T. D., Ludwig, J. L., & Phillips, K. (1995). Bringing order out of chaos: Psychometric characteristics of the confusion, hubbub, and order scale. *Journal of Applied Developmental Psychology, 16*(3), 429–444. https://doi.org/10.1016/0193-3973(95)90028-4

Maxwell, L. E. (1996). Multiple effects of home and day care crowding. *Environment and Behavior, 28*(4), 494–511. https://doi.org/10.1177/0013916596284004

Maxwell, L. E., Mitchell, M. R., & Evans, G. W. (2008). Effects of play equipment and loose parts on preschool children's outdoor play behavior: An observational study and design intervention. *Children, Youth and Environments, 18*(2), 36–63.

Mills-Koonce, W. R., Willoughby, M. T., Garrett-Peters, P., Wagner, N., Vernon-Feagans, L., & Family Life Project Key Investigators. (2016). The interplay among socioeconomic status, household chaos, and parenting in the prediction of child conduct problems and callous–unemotional behaviors. *Development and Psychopathology, 28*(3), 757–771. https://doi.org/10.1017/s0954579416000298

Mshida, H. A., Kassim, N., Mpolya, E., & Kimanya, M. (2018). Water, sanitation, and hygiene practices associated with nutritional status of under-five children in semi-pastoral communities Tanzania. *The American Journal of Tropical Medicine and Hygiene, 98*(5), 1242–1249. https://doi.org/10.4269/ajtmh.17-0399

Muñoz-Quezada, M. T., Lucero, B. A., Barr, D. B., Steenland, K., Levy, K., Ryan, P. B., Iglesias, V., Alvarado, S., Concha, C., Rojas, E., & Vega, C. (2013). Neurodevelopmental effects in children associated with exposure to organophosphate pesticides: A systematic review. *Neurotoxicology, 39*, 158–168. https://doi.org/10.1016/j.neuro.2013.09.003

Napier, M. A., Brown, B. B., Werner, C. M., & Gallimore, J. (2011). Walking to school: Community design and child and parent barriers. *Journal of Environmental Psychology, 31*(1), 45–51. https://doi.org/10.1016/j.jenvp.2010.04.005

Needleman, H. L., Schell, A., Bellinger, D., Leviton, A., & Allred, E. N. (1990). The long-term effects of exposure to low doses of lead in childhood: An 11-year follow-up report. *New England Journal of Medicine, 322*(2), 83–88. https://doi.org/10.1056/nejm199001113220203

Ng, S., Lin, C. C., Jeng, S. F., Hwang, Y. H., Hsieh, W. S., & Chen, P. C. (2014). Mercury, APOE, and child behavior. *Chemosphere, 120C*, 123–130. https://doi.org/10.1016/j.chemosphere.2014.06.003

Nsamenang, A. B. (1992). Human development in cultural context: A third world perspective. *Sage Publications*. https://doi.org/10.4135/9781483326030

Oishi, S. (2010). The psychology of residential mobility: Implications for self, social relationships, and well-being. *Perspectives on Psychological Science, 5*, 5–21. https://doi.org/10.1177/1745691609356781

Parcel, T. L., & Menaghan, E. G. (1993). Family social capital and children's behavior problems. *Social Psychology Quarterly, 56*(2), 120–135. https://doi.org/10.2307/2787001

Patel, N. B., Xu, Y., McCandless, L. C., Chen, A., Yolton, K., Braun, J., Jones, R. L., Dietrich, K. N., & Lanphear, B. P. (2019). Very low-level prenatal mercury exposure and behaviors in children: The HOME study. *Environmental Health, 18*(1), 1–12. https://doi.org/10.1186/s12940-018-0443-5

Pempek, T. A., & Lauricella, A. R. (2017). The effects of parent-child interaction and media use on cognitive development in infants, toddlers, and preschoolers. In F. C. Blumberg & P. J. Brooks (Eds.), *Cognitive development in digital contexts* (pp. 53–74). Academic Press. https://doi.org/10.1016/B978-0-12-809481-5.00003-1

Perera, F. P. (2017). Multiple threats to child health from fossil fuel combustion: Impacts of air pollution and climate change. *Environmental Health Perspectives, 125*(2), 141–148. https://doi.org/10.1289/EHP299

Perera, F. P., Wang, S., Rauh, V., Zhou, H., Stigter, L., Camann, D., Jedrychowski, W., Mroz, B., & Majewska, R. (2013). Prenatal exposure to air pollution, maternal psychological distress, and child behavior. *Pediatrics, 132*(5), e1284–e1294. https://doi.org/10.1542/peds.2012-3844

Prince, D. (2014). What about place? Considering the role of physical environment on youth imagining of future possible selves. *Journal of Youth Studies, 17*(6), 697–716. https://doi.org/10.1080/13676261.2013.836591

Repetti, R., & Wang, S. (2010). Parent employment and chaos in the family. In G. W. Evans & T. D. Wachs (Eds.), *Chaos and its influence on children's development* (pp. 191–208). American Psychological Association. https://doi.org/10.1037/12057-012

Repetti, R., Wang, S., & Saxbe, D. (2009). Bringing it all back home: How outside stressors shape families' everyday lives. *Current Directions in Psychological Science, 18*(2), 106–111. https://doi.org/10.1111/j.1467-8721.2009.01618.x

Riddell, A. (2008). *Factors influencing educational quality and effectiveness in developing countries: A review of research*. GTZ.

Robson, D. A., Allen, M. S., & Howard, S. J. (2020). Self-regulation in childhood as a predictor of future outcomes: A meta-analytic review. *Psychological Bulletin, 146*(4), 324–354. https://doi.org/10.1037/bul0000227

Roeters, A., Van Der Lippe, T., & Kluwer, E. S. (2010). Work characteristics and parent-child relationship quality: The mediating role of temporal involvement. *Journal of Marriage and Family, 72*(5), 1317–1328. https://doi.org/10.1111/j.1741-3737.2010.00767.x

Rohan, K. J., Roecklein, K. A., & Haaga, D. A. (2009). Biological and psychological mechanisms of seasonal affective disorder: A review and integration. *Current Psychiatry Reviews, 5*(1), 37–47. https://doi.org/10.2174/157340009787315299

Rollings, K., & Evans, G. W. (2019). Design moderators of perceived crowding and chronic physiological stress among children. *Environment and Behavior, 51*, 590–621. https://doi.org/10.1177/0013916518824631

Rosen, L. N., Targum, S. D., Terman, M., Bryant, M. J., Hoffman, H., Kasper, S. F., Hamovit, J. R., Docherty, J. P., Welch, B., & Rosenthal, N. E. (1990). Prevalence of seasonal affective disorder at four latitudes. *Psychiatry Research, 31*(2), 131–144. https://doi.org/10.1016/0165-1781(90)90116-m

Saegert, S. (1982). Environment and children's mental health: Residential density and low income children. In A. Baum, J. E. Singer, & J. L. Singer (Eds.), *Handbook of psychology and health: Issues in child health and adolescent health* (Vol. 2, pp. 247–271). Routledge. https://doi.org/10.4324/9780203781074

Saltzman, J. A., Musaad, S., Bost, K. K., McBride, B. A., & Fiese, B. H. (2019). Associations between father availability, mealtime distractions and routines, and maternal feeding responsiveness: An observational study. *Journal of Family Psychology, 33*(4), 465–475. https://doi.org/10.1037/fam0000519

Sampson, R. J., & Winter, A. S. (2018). Poisoned development: Assessing childhood lead exposure as a cause of crime in a birth cohort followed through adolescence. *Criminology, 56*(2), 269–301. https://doi.org/10.1111/1745-9125.12171

Searle, A. K., Baghurst, P. A., van Hooff, M., Sawyer, M. G., Sim, M. R., Galletly, C., Clark, L. S., & McFarlane, A. C. (2014). Tracing the long-term legacy of childhood lead exposure: A review of three decades of the Port Pirie cohort study. *Neurotoxicology, 43*, 46–56. https://doi.org/10.1016/j.neuro.2014.04.004

Serpell, R. (2011). Social responsibility as a dimension of intelligence, and as an educational goal: Insights from programmatic research in an African society. *Child Development Perspectives, 5*(2), 126–133. https://doi.org/10.1111/j.1750-8606.2011.00167.x

Shamama-tus-Sabah, S., & Gilani, N. (2010). Household chaos and its association with maternal education, family system, and children's academic achievement in Pakistani culture. *Pakistan Journal of Psychological Research, 25*(1), 19–30. https://doi.org/10.1186/s12889-020-08587-8

Shamama-tus-Sabah, S., Gilani, N., & Wachs, T. D. (2011). Relation of home chaos to cognitive performance and behavioral adjustment of Pakistani primary school children. *International Journal of Behavioral Development, 35*(6), 507–516. https://doi.org/10.1177/0165025411406852

Shih, R. A., Glass, T. A., Bandeen-Roche, K., Carlson, M. C., Bolla, K. I., Todd, A. C., & Schwartz, B. S. (2006). Environmental lead exposure and cognitive function in community-dwelling older adults. *Neurology, 67*(9), 1556–1562. https://doi.org/10.1212/01.wnl.0000239836.26142.c5

Sinharoy, S. S., Clasen, T., & Martorell, R. (2020). Air pollution and stunting: A missing link? *The Lancet Global Health, 8*(4), e472–e475. https://doi.org/10.1016/S2214-109X(20)30063-2

South, S. J., Haynie, D. L., & Bose, S. (2005). Residential mobility and the onset of adolescent sexual activity. *Journal of Marriage and Family, 67*(2), 499–514. https://www.jstor.org/stable/3600284

Steg, L. E., & De Groot, J. I. (Eds.). (2019). *Environmental psychology: An introduction* (2nd ed.). Wiley-Blackwell.

Swedo, S. E., Pleeter, J. D., Richter, D. M., Hoffman, C. L., Allen, A. J., Hamburger, S. D., Turner, E. H., Yamada, E. M., & Rosenthal, N. E. (1995). Rates of seasonal affective disorder in children and adolescents. *The American Journal of Psychiatry, 152*(7), 1016–1019. https://doi.org/10.1176/ajp.152.7.1016

Tappe, K. A., Glanz, K., Sallis, J. F., Zhou, C., & Saelens, B. E. (2013). Children's physical activity and parents' perception of the neighborhood environment: Neighborhood impact on kids study. *International Journal of Behavioral Nutrition and Physical Activity, 10*(1), 1–10. https://doi.org/10.1186/1479-5868-10-39

Taylor, M. P., Forbes, M. K., Opeskin, B., Parr, N., & Lanphear, B. P. (2016). The relationship between atmospheric lead emissions and aggressive crime: An ecological study. *Environmental Health, 15*(1), 1–10. https://doi.org/10.1186/s12940-016-0122-3

Tonetti, L., Barbato, G., Fabbri, M., Adan, A., & Natale, V. (2007). Mood seasonality: A cross-sectional study of subjects aged between 10 and 25 years. *Journal of Affective Disorders, 97*, 155–160. https://doi.org/10.1016/j.jad.2006.06.008

Uline, C., & Tschannen-Moran, M. (2008). The walls speak: The interplay of quality facilities, school climate, and student achievement. *Journal of Educational Administration, 46*, 55–73. https://doi.org/10.1108/09578230810849817

UNICEF. (2010). *Water, sanitation and hygiene.* Retrieved 20 November 2012 from http://www.unicef.org/media/media_45481.html

Van Dijk-Wesselius, J. E., Maas, J., Hovinga, D., Van Vugt, M., & Van den Berg, A. E. (2018). The impact of greening schoolyards on the appreciation, and physical, cognitive and social-emotional Well-being of schoolchildren: A prospective intervention study. *Landscape and Urban Planning, 180*, 15–26. https://doi.org/10.1016/j.landurbplan.2018.08.003

Van Maele-Fabry, G., Hoet, P., & Lison, D. (2013). Parental occupational exposure to pesticides as risk factor for brain tumors in children and young adults: A systematic review and meta-analysis. *Environment International, 56*, 19–31. https://doi.org/10.1016/j.envint.2013.02.011

Vernon-Feagans, L., Willoughby, M., & Garrett-Peters, P. (2016). Predictors of behavioral regulation in kindergarten: Household chaos, parenting, and early executive functions. *Developmental Psychology, 52*(3), 430–441. https://doi.org/10.1037/dev0000087

Villanueva, K., Badland, H., Kvalsvig, A., O'Connor, M., Christian, H., Woolcock, G., Giles-Corti, B., & Goldfeld, S. (2016). Can the neighborhood built environment make a difference in children's development? Building the research agenda to create evidence for place-based children's policy. *Academic Pediatrics, 16*(1), 10–19. https://doi.org/10.1016/j.acap.2015.09.006

von Ahlefeld, H. (2007). What the OECD PISA study reveals about the physical learning environment. In E. Knapp, K. Noshis, & C. Pasalar (Eds.), *School building design and learning performance* (pp. 35–40). Comportements.

Wachs, T. D., & Camli, O. (1991). Do ecological or individual characteristics mediate the influence of the physical environment upon maternal behavior. *Journal of Environmental Psychology, 11*(3), 249–264. https://doi.org/10.1016/S0272-4944(05)80186-0

Wachs, T. D., & Corapci, F. (2003). Environmental chaos, development and parenting across cultures. In C. Raeff & J. Benson (Eds.), *Social and cognitive development in the context of individual, social and cultural processes* (pp. 54–83). Routledge.

Weisner, T. S. (2002). Ecocultural understanding of children's developmental pathways. *Human Development, 45*(4), 275–281. https://doi.org/10.1159/000064989

Weisner, T. S. (2010). Well-being, chaos and culture: Sustaining a meaningful daily routine. In G. W. Evans & T. D. Wachs (Eds.), *Chaos and its influence on children's development* (pp. 211–224). American Psychological Association.

Wells, N. M. (2000). At home with nature: Effects of "greenness" on children's cognitive functioning. *Environment and Behavior, 32*(6), 775–795. https://doi.org/10.1177/00139160021972793

Wells, N. M., & Evans, G. W. (2003). Nearby nature: A buffer of life stress among rural children. *Environment and Behavior, 35*(3), 311–330. https://doi.org/10.1177/0013916503035003001

Wells, N. M., & Harris, J. D. (2007). Housing quality, psychological distress, and the mediating role of social withdrawal: A longitudinal study of low-income women. *Journal of Environmental Psychology, 27*(1), 69–78. https://doi.org/10.1016/j.jenvp.2006.11.002

Wells, N. M., & Moch, A. (2003). Housing and mental health: A review of the evidence and a methodological and conceptual critique. *Journal of Social Issues, 59*(3), 475–500.

Wener, R. E., & Evans, G. W. (2011). Transportation and health: The impact of commuting. In J. Nriagu (Ed.), *Encyclopedia of environmental health* (pp. 400–407). Elsevier. https://doi.org/10.1016/B978-0-444-52272-6.00703-0

Werner, C. D., Linting, M., Vermeer, H. J., & Van IJzendoorn, M. H. (2015). Noise in center-based child care: Associations with quality of care and child emotional wellbeing. *Journal of Environmental Psychology, 42*, 190–201. https://doi.org/10.1016/j.jenvp.2015.05.003

Whipple, S. S., & Evans, G. W. (2022). Environmental psychology and children's social development. In P. K. Smith & C. Hart (Eds.), *The Wiley-Blackwell handbook of childhood social development* (3rd ed.). Wiley-Blackwell.

Whitbeck, L. B., Simons, R. L., Conger, R. D., Wickrama, K. A. S., Ackley, K. A., & Elder, G. H. (1997). The effects of parents' working conditions and family economic hardship on parenting behaviors and children's self-efficacy. *Social Psychology Quarterly, 60*(4), 291–303. https://doi.org/10.2307/2787091

WHO/UNICEF Joint Monitoring Program for Water Supply, Sanitation, & Hygiene (JMP). (2021). *Progress on household drinking water, sanitation and hygiene 2000–2020: Five years into the SDGs*.

Wigle, D. T. (2003). *Child health and the environment*. Oxford University Press. Child Health and the Environment - Donald T. Wigle - Oxford University Press (oup.com)

Wilkie, S., Townshend, T., Thompson, E., & Ling, J. (2018). Restructuring the built environment to change adult health behaviors: A scoping review integrated with behavior change frameworks. *Cities & Health, 2*(2), 198–211. https://doi.org/10.1080/23748834.2019.1574954

Xu, J., Hu, H., Wright, R., Sánchez, B. N., Schnaas, L., Bellinger, D. C., Park, S. K., Martínez, S., Hernández-Avila, T.-R., & Wright, R. O. (2015). Prenatal lead exposure modifies the impact of maternal self-esteem on children's inattention behavior. *The Journal of Pediatrics, 167*(2), 435–441. https://doi.org/10.1016/j.jpeds.2015.04.057

Yogman, M., Garner, A., Hutchinson, J., Hirsh-Pasek, K., Golinkoff, R. M., Committee on Psychosocial Aspects of Child and Family Health, & Council on Communications and Media (2018). The power of play: A pediatric role in enhancing development in young children. Pediatrics, 142(3) e20182058. https://doi.org/10.1542/peds.2018-2058.

Yorifuji, T., Kashima, S., Higa Diez, M., Kado, Y., Sanada, S., & Doi, H. (2016). Prenatal exposure to traffic-related air pollution and child behavioral development milestone delays in Japan. *Epidemiology, 27*(1), 57–65. https://doi.org/10.1097/ede.0000000000000361

Chapter 6
Equitable Change in Community Built Environments for Family Health: REACH River Rouge Project

Laurie Lachance, Samuel Shopinski, Theresa Tejada, Herman Jenkins, and Martha Quinn

The health of children and families is markedly influenced by conditions in the environment, which affect opportunities for healthy behaviors such as engaging in physical activity (Centers for Disease Control and Prevention [CDC], 2018; Fielding et al., 2010; Gebbie et al., 2003; Woolf & Aron, 2013). Opportunities for physical activity in communities are affected by economic and social conditions such as housing stability, food security, safety, access to transportation, and infrastructures in neighborhoods (House et al., 1994; Larson et al., 2009; Lovasi et al., 2009; National Academies of Sciences, Engineering, and Medicine [NASEM], 2016; Williams, 1990, 1997). Poor economic and social conditions restrict and constrain opportunities for healthy choices. In communities that experience inequities, these conditions are often the result of disinvestment and a lack of resources ensuing from racial discrimination (Borrell et al., 2006; Krieger, 2014; Lewis et al., 2015; Williams & Mohammed, 2009). Awareness of established systems and policies and consideration of how they operate in the context of a community (Waters Center for Systems Thinking, 2021; Yu, 2020; Zurcher et al., 2018b) enables capacity building for community-driven change to social and economic conditions. Capacity building refers to the ability of an entity to effectively achieve its mission now and to sustain it in the future.

L. Lachance (✉) · M. Quinn
School of Public Health, Department of Health Behavior and Health Education, The University of Michigan, Ann Arbor, MI, USA
e-mail: lauriel@umich.edu

S. Shopinski · T. Tejada
The National Kidney Foundation of Michigan, Ann Arbor, MI, USA

H. Jenkins
Friends of the Rouge, Plymouth, MI, USA

S. E. Ortiz et al. (eds.), *Environmental Impacts on Families*, National Symposium on Family Issues 12, https://doi.org/10.1007/978-3-031-22649-6_6

It is through a lens of equity that the following work related to physical activity, environment, and family health is presented. This chapter draws on the work of the Food & Fitness (F & F) community partnerships (Lachance et al., 2014a; Lachance et al. 2018a, c), an initiative funded by the W.K. Kellogg Foundation (WKKF) and established to create community-determined change in conditions that affect opportunities for health and health equity in neighborhoods across the USA (Lachance et al., 2014b; Zurcher et al., 2018a;). The focus of the F & F partnerships has been to affect opportunities for health by increasing access to locally grown food and safe places for physical activity in communities with inequities by changing policies, infrastructure, and systems at a local level (Lachance, Quinn, et al., 2018).

This chapter includes a case study of an initiative currently underway in Southeast Michigan. The Racial and Ethnic Approaches to Community Health Program (REACH; NCCDPHP, 2020), funded by the Centers for Disease Control and Prevention (CDC) and led by the National Kidney Foundation of Michigan (NKFM), focuses on strategies to improve physical activity and the built environment, among other conditions, in order to make healthy choices easier. Recipients of REACH grants routinely engage with the priority population throughout program design, implementation, and evaluation; hire or partner with individuals and organizations that can represent the needs of the community; and adapt their programs to reflect cultural norms, language, and practices.

In 2018, the NKFM was one of 31 grantees awarded funding through the REACH program. The NKFM is a not-for-profit organization whose mission is to prevent kidney disease and improve the quality of life for those living with it, focusing on vulnerable populations where chronic kidney disease and its leading causes – diabetes and hypertension – are prevalent. In their REACH initiative, the NKFM aims to improve health outcomes, prevent chronic disease, and reduce health disparities among African Americans and Hispanics, populations with the highest risk or burden of chronic disease in the neighboring cities of Inkster, Wayne, and Westland, Michigan.

Elements of Change for Equitable Active Living/Built Environments

The process of creating opportunities for physical activity in communities involves operating through a lens of equity, using a community-driven systems thinking approach, assessing the current environment, identifying and prioritizing potential efforts, and including evaluation and dissemination of community changes (Fig. 6.1).

Each of these elements will be addressed in relation to the active living/built environment. The case study of the Lower River Rouge project is presented after these sections and illustrates these important elements of community change.

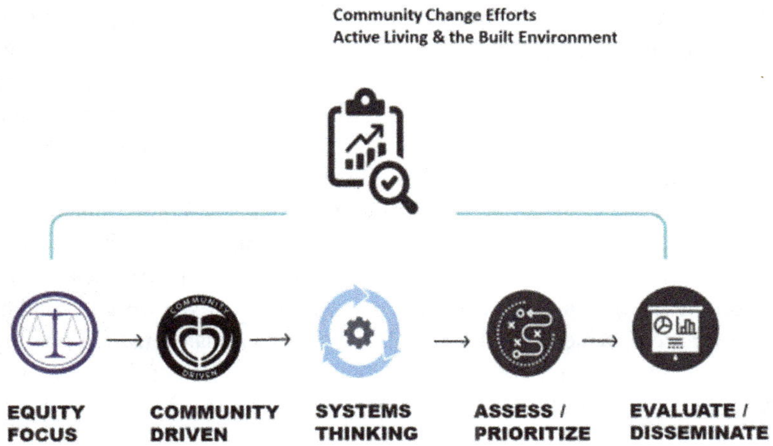

Fig. 6.1 Elements of community change efforts to create equitable active living/built environments

Focus on Equity

Equity refers to assumptions, practices, and institutional behaviors that reform the social structures that consistently consign too few benefits and too many obstacles for health to those in vulnerable neighborhoods, especially people of color with low income (Institute for Democratic Renewal and Project Change Anti-Racism Initiative, 2001). Efforts to improve health equity are influenced by the complex social and physical environments where vulnerable families and children live, work, and play. For example, there is less availability of safe places for physical activity in low-income neighborhoods, particularly low-income neighborhoods of color (Auchincloss et al., 2009; Franco et al., 2008; Landrine & Corral, 2009).

Individual health behaviors are shaped by systemic, avoidable, unjust, community-wide inequities in access that in turn affect health status and the distribution of morbidity and mortality across neighborhoods (Braveman et al., 2017; Epping-Jordan et al., 2004; Frieden et al., 2010; Schmid et al., 1995; Story et al., 2008). Improvements to health in communities are dependent on narrowing the equity gap and focusing on justice (Clark et al., 2011; Israel et al., 2010; Livingood et al., 2011).

Health equity is about inclusive access to safe places and is not always about everyone getting the same thing – fair is not equal when considering equitable opportunities. It is important to consider who will benefit and who will be harmed when strategizing changes and prioritizing efforts. Placing those who are most impacted by potential changes at the center of conversations will increase the potential for finding the best solutions to problems facing neighborhoods and families.

Efforts to improve health equity should focus on both the population and the gaps in opportunities for health. Efforts to change conditions at the population level that focus only on the population as a whole can result in shifting health-related risk, while the gap in opportunities for health at the individual level does not change. In some instances, the gap at the individual level has been shown to increase due to variability in baseline risk and resources (Frohlich & Potvin, 2008; Victora et al., 2001). For example, creating facilities in communities that require transportation or membership fees can disproportionately create advantages to those with more resources. Strategies that address a health equity gap, while at the same time increasing opportunities for health in the overall population, move the work beyond merely a focus on access, which is only one of many factors that constrain healthy outcomes (Frohlich & Potvin, 2008; Powell, 1999). We cannot consider our work successful if we leave people behind.

Community-Driven Change Process

Meaningful engagement of community members is a primary goal of local systems and policy changes aimed at increasing families' opportunities for physical activity. Community-driven approaches to influence the complex social and physical environments that shape opportunities for health can strengthen the ability of neighborhood residents, organizations, and institutions to foster and sustain neighborhood change (DeFilippis, 2001; Kubisch et al., 2010; Schulz et al., 2013). Grassroots organizations, nonprofit and for-profit organizations, government agencies, and citizens living and working in the community are not always operating in relation to each other. A first step toward community change involves building relationships among all of these groups and realizing a common vision for active living in the community. Part of this step is the realization that often these groups do not have equal footing and decision-making power. Therefore, a conscious effort is needed to create mechanisms for sharing of power and resources.

With respect to efforts to change a community's opportunities for health, capacity building should allow for citizens to take on leadership roles in their communities and inform decision-making. In order to accomplish these changes, partnerships need to recruit and support community members as leaders, equip partnership members with advocacy and policy change skills, ensure that those who are in leadership positions reflect the community's positions, monitor resources, and recruit strategic partners who can further influence the decision-making to create the needed changes.

Building capacity within the context of community change is a democratic process that provides residents with the tools to be agents of their own change (Clark et al., 2006). These efforts create opportunities for community members to be included in discussions and decisions about how change efforts should be accomplished. Efforts that are led by community members underscore the values of equity, self-determination, social justice, and respect for diversity that are fundamental to healthy communities (Clark et al., 2006; DeFilippis, 2001; Kubisch et al., 2010).

Everyone involved in community change initiatives – funders, leaders, community members, and organizational members – needs to be prepared to address the discomfort that can result from strengthening grassroots efforts and shifting the current power structure. Funders are sometimes interested in scaling community change work. They may aspire to increase the number of changes within a community or expand the work throughout a region, state, or nation. Yet, there is tension between this desire to broaden impact and the desire to address health equity, which necessitates greater local focus on community context. Balancing funder and grantee goals with community capacity and needs enables more effective investments. Balancing goals and capacity also require guarding against behaviors that are driven by a need for funding over consideration of the community's needs and desires.

Systems Thinking Approach

Over the last few decades, the understanding of inequities related to health in populations has shifted from a focus on individuals and individual-level risk factors (Keys, 1980; Nieto, 1999) to the realization that risk is generated by conditions that influence opportunity for health (House et al., 2008; Lantz et al., 1998; Williams & Collins, 2001) and that these conditions operate at the level of neighborhoods and communities (Diez Roux, 2007; Morenoff & Lynch, 2004; Schulz et al., 2012). Thus, a systems approach, or focusing on systems, can make healthier choices into easier choices for families and children.

Systems thinking approaches allow for efforts that are owned and led by communities through engagement of individuals with divergent experience, expertise, and perspectives and alignment of a vision that respects the unique culture, context, and assets of those communities (Zurcher et al., 2018b). Systems thinking approaches seek to understand a system as a whole, focusing on causal relationships among parts of a system rather than on the parts themselves, and examining the system from multiple perspectives (Zurcher et al., 2018b; Waters Center for Systems Thinking, 2021).

Establishing systems thinking as a framework includes realizing that relationships are primary and community ownership is essential. Collective thinking about how organizations and resources fit together and time taken to create a shared vision result in more robust planning and are foundational for change efforts. In systems thinking approaches, there is an understanding of slower being faster (Kim, 2001). If time is not taken to build these foundations and relationships, much of the work that is done will later need to be undone and at a cost to resources, trust, and opportunity.

Thinking in terms of sustainable change within systems takes practice. When faced with complex problems, we are often influenced by society's messages about what constitutes enough time. Systems thinking invites us to slow down and allow for alignment with the system we are attempting to influence. Oftentimes those who have been working to create healthier communities focus on programs and practices even when they have set out to create systems change. Although programs are needed in systems change, it is important that they are linked with change efforts and that there is careful consideration of the role programs play in changing systems (Zurcher et al., 2018b).

Numerous studies have identified targeted local policy change as a promising strategy for creating systems change in infrastructure, services, and activities that can yield population-wide improvements in physical activity and healthful outcomes (Brownson et al., 2006; Frieden et al., 2010; Papas et al., 2007; Sacks et al., 2009). The accessibility of safe and inviting spaces for physical activity depends on the number and diversity of these spaces; their location relative to where people live, work, or go to school; and the quality and safety of connections and routes to destinations (Lachance et al., 2018a, b, d). The quality and safety of spaces for physical activity are based on the appropriateness and good repair of the facilities; aesthetic features such as good design, greenery, and cleanliness; social factors such as absence of crime; and perceptions of vibrancy. The spaces can be either built or natural. Accessibility is further determined by programmatic supports in these spaces such as recreation or physical education programs, walking/active commuting programs, and ability to utilize these programs (e.g., cost, staffing, hours of operation). With a vision in place, community partnerships can gather both existing and new data to provide the most complete picture of the current community environment as possible.

Assessment of the Built Environment and Prioritization of Potential Efforts

The assessment process is valuable for community engagement at all levels, for education of partners and community members, and as a potential community organizing method. The key purpose is to inform the development of the most promising strategic options and ultimately the strategic priorities that form the efforts that will be undertaken to improve the built environment.

Assessments identify opportunities and potential action strategies for planning and prioritization of efforts; address community and partner preferences and perceptions; and explore what constituents of the community experience, believe, care about, and are willing to work on. Methods for community assessment include focus groups, formal and informal community meetings, surveys, previous studies, and one-on-one discussions. Assessments of the built environment investigate the physical conditions at both the neighborhood and citywide scale and how conditions can be improved through walking audits, GIS data, Photovoice, documentation of current conditions, and secondary data from local agencies. Locating previous reports and assessments often involves tracking down information held by various stakeholders and decision-maker individuals and organizations within the community, and these sources can differ across communities.

A review of existing local policies and plans to identify opportunities for development is also part of the assessment process. Local systems are most often shaped by various local policies related to organizational leadership, decision-making, administrative structures and rules, operating procedures, design guidelines, budgetary practices and priorities, staff training, professional norms and attitudes, and levels of coordination with other related departments or units (often referred to as little "p" policy in relation to legislative changes or big "P" policy). Local policies can be uncovered through stakeholder interviews, policy audits/analyses, and discussions with informal leaders.

Community assessments of the active living/built environment can consider three domains of the built environment: (1) active transportation, (2) parks and recreation facilities, and (3) school systems. All three of these domains are influenced by social and environmental conditions beyond the built environment such as crime, climate, pollution, language, cultural beliefs and practices, race, and poverty. From a systems perspective, these domains often intersect, and when conducting assessments, it can be helpful to think of them as distinct parts of overall systems that can be examined both separately and together. *Active transportation* pertains to the physical arrangements and facilities that support walking, biking, public transit, and other active means of transportation. The domain of *parks and recreation* refers to the accessibility and quality of parks, playgrounds, trails and natural open space, indoor recreation centers, gyms, pools, and the programs that encourage people to use them. *School systems* represent school environments including buildings, grounds, curricula, procedures, and norms, as well as the accessibility of schools and the quality of routes to and from school.

Assessment and Prioritization of Efforts Related to Active Transportation

Active transportation includes walking, biking, public transit, and other active means of mobility and can be a part of routine physical activity. A community's transportation system determines the accessibility, safety, comfort, and feasibility for traveling actively and impacts not only how people move from place to place but also the fundamental character of communities and the choices and opportunities that are available.

Assessment of current local policies and systems and prioritization of efforts related to active transportation includes having partners and a process in place to address active living capital investments, connectivity/integrated networks, safety and speed limits, available facilities, implementation of existing plans, aesthetics, incentives and public campaigns, information and communication, training and education, and targets for behaviors, safety, and data collection (see Table 6.1). Overall assessment of the built environment should address whether pedestrian/bicycling facilities (e.g., sidewalks, bike lanes, crosswalks and signals, benches, racks, design amenities) are available on major transportation routes between centers of activity such as residential neighborhoods, employment and shopping centers, schools, parks, libraries, groceries, and farmers' markets. Availability assessments include considerations of where these facilities are most needed to ensure safety and comfort for pedestrians and cyclists, such as wide or high-speed roads, areas with high likelihood of potential conflicts with motor vehicles. Ideally there should be dedicated pathways that are separated from traffic and that are part of a larger, integrated, and signed pedestrian/bike network that connects a variety of major destinations in a convenient way. Street design, speed limits, and traffic law enforcement should ensure safety for all users, provide comfort for pedestrians and cyclists, and be equitably distributed across all neighborhoods and for all types of users. Active transportation should be well integrated with quality public transportation service and combined with street amenities that pedestrians and cyclists find attractive, including trees/landscaping, art, benches, cafes, lighting, and signage.

Assessment and Prioritization of Efforts Related to Parks and Recreation

The built environment domain of parks and recreation refers to the accessibility and quality of parks, trails, playgrounds, natural open space, indoor recreation centers, pools, gyms, and the programs that encourage people to use these facilities. Not all families and children enjoy equal access to parks or indoor recreation facilities. When available and affordable, indoor recreation facilities provide physical activity opportunities throughout the year, regardless of the weather, and can concentrate amenities in one place.

One way to help address the availability of parks, trails, and greenways is to create and use standards for park creation, development, and management with

Table 6.1 Questions for assessment and prioritization of active transportation efforts

	Active transportation	
Overall questions	1. How up-to-date and well integrated are the community's pedestrian, bicycling, and transit plans in relation to other city priorities and plans? 2. How could current and planned active transportation policies and procedures be improved and applied more equitably? 3. What, if any, new capacities need to be built within the transportation system to develop this knowledge and help ensure delivery of quality transportation opportunities?	
Current policies and procedures	What policies or procedures does local government currently have in place to create opportunities for active transportation in the community?	
	Assessment categories	**Examples of efforts**
	Partners/process in place to address active living	A multidisciplinary committee/board to advise the mayor or city council about opportunities for improving active transportation
	Capital investments	Capital investments consistent with and sufficient to meet active transportation goals and priorities
	Connectivity/integrated network	Require connectivity for new streets/neighborhoods; intermodal travel between public transit, walking, and bicycling (e.g., bike racks on buses, access to public transportation and facilities); safely integrate greenways; initiatives to complete streets include specific, measurable standards
	Safety/speed limits	Address design speeds of roads, speed limits and signage; link investments to areas of primary safety risk; lower vehicle speeds in neighborhoods and near popular destinations or centers of activity
	Facilities	Equitably distribute ped/bike/transit-friendly facilities in all neighborhoods for all types of users
	Implementation of existing plans	City policies, plans, codes, and programs implemented to create a more pedestrian- and bicycle-friendly community; city pedestrian/bicycle coordinator position funded
	Aesthetics	Amenities and features that make streets more attractive to pedestrians/cyclists (e.g., benches, lighting, trees and landscaping, water fountains, public art, directional signage, trash cans)
	Incentives/public campaigns	Requirements/incentives for transportation demand management (TDM) measures (e.g., free transit passes to new residents and employees; funding for parking district or transportation improvement district; instead of building parking, traffic reduction programs; parking cash-out for employees); encourage walking/biking in areas where significant segments of the population do not drive and where short trips are most common

(continued)

Table 6.1 (continued)

	Active transportation	
	Information/ communication	Information programs to promote active transportation for all purposes and to communicate the many benefits of active transportation to residents and businesses (e.g., with maps, campaigns, neighborhood walks/ rides, or a walk/ride with the mayor)
	Training/education	Train/encourage transportation engineers to better design facilities and enable active transportation (e.g., context-sensitive design, complete streets); educate road users to share the road and interact safely
	Targets for behaviors/safety/ data collection	Adopt a target level of walking or bicycle use (e.g., percent of trips) and safety to be achieved within a specific timeframe, and improve data collection necessary to monitor progress

measures such as minimum acres per number of residents, actual benefits to residents, property value improvements, and other measurable outcomes. Other policy strategies include integrating parks into the redevelopment of low-income neighborhoods with large numbers of vacant lots and brownfields; using developer impact fees to create and maintain parks, trails, and greenways; and leveraging private and public financing for park creation and maintenance.

The design and staffing of parks, trails, greenways, and indoor facilities can encourage or inhibit physical activity in these spaces. Reduction in park budgets has contributed to the neglect of the built environment, leaving fields and equipment in disrepair. Physical features such as restroom facilities, drinking fountains, and shade may be absent, in a dilapidated state, or inaccessible to persons living with disabilities.

Policy strategies to address these shortcomings include increasing community involvement in the advocacy for park funding and in the maintenance of parks, trails, and greenways using the "Friends of" model for advocacy groups and establishing joint use agreements with local schools for cost-sharing purposes. Design features such as connectivity between common local and regional destinations and the selection of surface materials that minimize maintenance needs and trip hazards and maximize usability are built environment strategies that can increase the use of trails and greenways for physical activity.

Safety concerns and fear of crime can limit the use of local parks, trails, and greenways. After years of neglect and reduction in staffing, many parks, trails, and greenways attract crime and violence. The design of spaces can obscure sight lines and create spaces where criminals can hide, contributing to a general sense of being unsafe. Design solutions that enhance visibility, such as improved lighting and maintenance that keeps vegetation low, coupled with supervision from paid and volunteer personnel, security cameras, emergency phones, and police presence, are policy strategies to address safety concerns. The number and type of programs available at local parks, greenways, and trails can influence their use for physical activity. Individual preferences related to age, race/ethnicity, gender, and exercise habits also influence use.

Assessment of current local policies and systems and prioritization of efforts related to parks and recreation facilities includes consideration of equitable access, connectivity, user needs, assets and resources, joint use and cost sharing, maintenance, safety, liability, and community involvement (see Table 6.2).

Assessment and Prioritization of Efforts Related to School Systems

The domain of school systems represents school environments including buildings, grounds, curricula, procedures, and norms, as well as the accessibility of schools (i.e., the quality of routes to and from school). The local policies that govern site selection, curriculum, and off-hours use of school facilities can either encourage or inhibit physical activity among children. Active transportation for school children includes walking, biking, and small-wheeled transport such as skating, skateboarding, or push scooters. The distance between home and school is often a barrier to using active transportation to and from school. As larger schools are built away from neighborhoods to accommodate more students, fewer children live close enough to routinely walk or bike to school. Policies that alter the school site selection process and encourage smaller neighborhood schools and equitable use of resources for those schools can increase the opportunity for active transportation.

Traffic danger is commonly reported as a reason for not engaging in active transportation to and from school. Heavy automotive traffic and the absence of sidewalks, crosswalks, bike lanes, and other safe routes increase students' risks. Other limiting factors include the presence of violence and crime within the community or the fear of crimes against children such as abduction or gang activity. Built environment changes may include street-level improvements such as the construction of sidewalks, bike paths, and trails, marked and raised crosswalks, additional or modified street signage including flashing speed limit signs, and traffic-calming measures including roundabouts, medians, and curb extensions.

Policy strategies aimed at improving active transportation include enforcing speed limits, reconsidering school policies that restrict walking and biking to school, locating crossing guards at busy or dangerous intersections, improving street and sidewalk lighting, and establishing neighborhood watch programs. Many communities have addressed both traffic danger and concerns about crime by initiating a Walking School Bus Program (CDC, 2005) where adult volunteers walk a small group of students to school along a designated route with set pickup locations.

During the school day, physical education classes and recess provide opportunities for physical activity. Currently, though, physical education in many US schools is deficient in quantity and quality, particularly in low wealth schools (Acosta et al., 2021). In general, school administrators are not held accountable for physical education under the No Child Left Behind legislation (No Child Left Behind Act, 2002). Administrators and teachers place little or no academic value on physical education,

Table 6.2 Questions for assessment and prioritization of parks and recreation facilities' efforts

	Parks and recreation facilities	
Overall questions	1. How up-to-date and well integrated is the community's park and recreation plan or trails and greenways master plan in relation to other priorities and plans? 2. How could current and planned policies and procedures for the community parks and recreation facilities be improved and applied more equitably? 3. What, if any, new capacities need to be built within the parks and recreation system to help ensure delivery of quality parks and recreation opportunities?	
Current policies and procedures	What policies or procedures does local government currently have in place to create opportunities for quality parks and recreation facilities in the community?	
	Assessment categories	**Examples of efforts**
	Equitable access	Increase percentage of residents living less than one-quarter mile from a major park, trail, greenway, or indoor recreation facility (if rural, less than a 15-minute drive); redevelopment policies encourage development of parks and recreation facilities or use them as a catalyst for new development; lower financial barriers to low-income individuals/families
	Connectivity	Safe connections to parks, trails, greenways, and indoor recreation facilities by walking, biking, or the use of public transit (e.g., ped/bike facilities, well-designed bus routes and waiting areas, subdivision easements)
	User needs	Recreation facilities and programs meeting needs and preferences of current users; facilities and programs monitored for underuse or oversubscribed; monitor who uses the facilities and programs and for what purpose
	Assets and resources	Assets and resources meet goals; leverage private resources or partner with voluntary and nonprofit organizations (e.g., "friends" groups, conservancies); balance budget between operating and capital funds; monitor where resources come from for facilities, programs, maintenance and security, and how expandable or renewable they are; measure and communicate outcomes and cost-efficiency to the public and its decision-makers
	Joint use/cost sharing	Manage land, facilities, or programs in joint use with a school district; make arrangements for sharing the costs of utilities, maintenance, security, janitorial, equipment, etc.
	Maintenance	Meet routine maintenance responsibilities for existing and planned facilities; create additional resources or methods for maintenance
	Safety	Provide safety and security for existing and planned facilities and programs; collect data on crime at or around sites
	Liability	Overcome liability concerns related to the use of the facilities

(continued)

Table 6.2 (continued)

	Parks and recreation facilities	
	Community involvement	Strengthen community involvement in the system; increase awareness of how responsive the system is to community feedback

especially when school funding is tied to academic performance and thus creates incentives for keeping students in their seats and focused on studies. These current policies allow many schools to fall below national guidelines for the amount of physical education offered to students, and there is little monitoring and enforcement even when requirements are in place. Certain states allow exemptions to national guidelines, so many high school students may not take any physical education. Budget constraints have contributed to the reduction or elimination of physical education classes, and personnel deficiencies limit both the variety of courses available and the total number of classes offered each week. Additionally, many schools have reduced or eliminated recess for elementary school students, limiting another opportunity for students to be physically active.

Physical education also suffers from poor or inconsistent quality. Many students do not spend the recommended minutes in physical education class (150/week for elementary, 225/week for secondary); classes are too large for students to achieve and maintain the recommended level of sustained vigorous activity; and the classes often lack quality playing fields, indoor space, and equipment. The shortage of qualified, credentialed physical education teachers and limited opportunities for professional development offered to existing teachers also affects the quality of classes. Policy changes to address the quantity and quality deficiencies observed in physical education include improvements to the monitoring and enforcement of state physical education requirements, employment of certified physical education teachers, adoption of activity-focused curricula, provision of professional development opportunities for teachers, and elimination of the use or withholding of physical activity as a punishment.

Given that the majority of physical activity among young people occurs outside of school hours, developmentally appropriate after-school programs are an important strategy for increasing physical activity. There are, however, a number of barriers to the availability, accessibility, and desirability of after-school programs. Inadequate resources affect the variety of activities offered and contribute to inadequately trained sports and recreation staff. Budget constraints also limit staffing and facility hours after school, during weekends and breaks. Budget cuts in many school districts result in a "pay-to-play" system that can limit participation by lower-income youth. Finally, many programs and agencies emphasize competitive after-school sports and intramurals, which can be a barrier for students with lower skill levels, who are physically unfit, or who may not be attracted to team sports.

Collaboration between schools and other agencies can begin to address these barriers to increasing physical activity in the domain of school systems. Joint use agreements encourage the development, maintenance, and use of facilities by schools and the surrounding communities. Collaboration between schools and the community can result in the use of community resources to provide equipment through sponsorship, the coordination of staffing, and access to facilities. Other strategies aimed at after-school programs should address the provision of a variety of competitive and noncompetitive sports and recreation, safe and adequate equipment and instruction, amenities such as cool water and shade, protection from violence and exposure to environmental hazards, and staff training in injury prevention, first aid, and skill development. Assessment of current local policies and systems and prioritization of efforts related to school systems includes consideration of transportation to and from school/active transport, active play, physical education, joint use agreements/shared use of facilities, and accessibility (see Table 6.3).

Table 6.3 Questions for assessment and prioritization of school systems' efforts

	School systems	
Overall questions	1. What outdoor and indoor physical facilities are available at and around schools to support safe places for physical activity and play? 2. How could current and planned policies and practices be improved? 3. How can capacity building within the school system develop this knowledge?	
Current policies and procedures	What policies or procedures do schools have in place to create opportunities for safe physical activity and play at and around schools?	
	Assessment categories	Examples of efforts
	Transportation to and from school/active transport	Encourage active transportation to and from school/safe routes to schools programs (e.g., sidewalks, bike racks, crosswalks, signals, crossing guards, bike lanes, greenways); provide transportation to home for students who participate in after-school intramural activities or physical activity clubs
	Active play	Provide safe equipment, a safe playground, and safe indoor facilities for active play; have daily recess breaks/time for a variety of physical activity options including unstructured play; train staff to lead activity sessions and model positive activity behaviors; have clean sources of tap water and working water fountains
	Physical education	Meet or exceed requirements for minimum minutes of physical education, trained physical educators, and various physical activity options that reflect the interests and diversity of program attendees (ranging from competitive sports to dance and individual fitness activities)
	Joint use agreements/ shared use of facilities	Make facilities available to children, adolescents, and adults outside of school hours or when school is not in session for physical activity programs (e.g., before school, after school, evenings, weekends, or school vacation)
	Accessibility	Adjust or waive fee structures for students who cannot afford to pay for physical activity programs

Assessment and Prioritization of Efforts Related to Policy Change

The assessment process will likely generate more potential local policy and systems strategies than any partnership will have the time, resources, and energy to address at once. A critical step toward changing the built environment in communities involves consideration of proposed systems and policy change efforts and selection of the most promising strategies; start smaller and look for ways to scale up and spread efforts over time as appropriate.

The best way to begin the process of prioritization is using the information gathered during the assessment process to become aware of existing assets in the community, understanding the timing and capacity for change, and enlisting the help of those most knowledgeable about local policy change. There is a danger that this process of identifying neighborhoods of focus and potential community partners, and assessing current conditions in a community, can become a lengthy one that can postpone or stall change efforts. However, defining the scope of work that is agreed upon and being ready for implementation is essential. Not taking time to build this strong foundation will work against efforts in the long run. The prioritization process should include a check for consistency with the vision and an examination of the ways in which the prioritized efforts can further refine the vision.

Prioritization efforts are most effective when there are multiple opportunities for participation by community partners and when the following considerations related to community impact, feasibility, and health equity of proposals are addressed:

- Benefits and costs to residents – For example, improved parks and streetscapes may increase the displacement of low-income households through gentrification; the expenditure of political and social capital or a damaged political relationship could diminish the likelihood or effectiveness of future attempts.
- Potential effectiveness – Logical and likely sequence of steps that lead to the intended outcome and supported by local experience, promising strategies from other communities, or research.
- Efficient use of resources – Whether there are alternatives to the proposed strategy that are viable pathways to the same, or better, outcome.
- Sustainability – Future outlook of a proposed policy and systems change; what types of monitoring and maintenance will be needed to ensure continued success?
- Viability related to the community context; political context, for example, indifference or opposition of key players, change of personnel/leadership, or loss of a champion; structural barriers related to existing rules, incentive, norms, habits, or resources can also be present.
- Available resources – How will resources be leveraged to achieve the goals?
- Readiness of decision-makers – Is there sufficient passion and preparation among partners to pursue this effort?
- Capacity related to relationships – Does the partnership have the particular capacities needed for long-term success of this strategy? Do healthy relationships and commitments exist within the collaborative that can contribute to a

successful outcome for this effort? Do the priorities of existing leadership support this effort?

- Commitment and skills needed to succeed – Community organizing experience, skilled facilitators or trainers, track record of effective advocacy, survey and evaluation expertise, long-standing community ties, fundraising skills, and engagement of partners already working on these issues.
- Advocacy of resident opinions – Create multiple opportunities for community members and leaders to advocate for their interests.
- Health inequities addressed – Does everyone in the community have a fair and just opportunity to be healthier? Do partners agree that a particular strategy can address health inequities and impact community members?

Understanding and addressing these considerations is a critical step before launching a potential effort and may help partners determine whether and when to pursue specific strategies. Structural barriers related to existing rules, incentive, norms, habits, or resources can be present; understanding and addressing any barriers is also critical before launching a potential effort. While it may be important to select policy/systems change strategies with broad appeal that can mobilize and maintain a vibrant coalition, strategies that do not also increase health equity may fall outside of the initiative's key priorities. An initial step for filtering potential strategies is to explicitly address health equity.

Evaluation of Change Efforts and Dissemination of Findings

Guiding principles for evaluating community change efforts for active living/built environments include ensuring that efforts are participatory, build capacity within the community, create systems and policy changes with outcomes that are sustainable, and focus on equity. Both explicit and implicit considerations of equity are a critical consideration for evaluation. This includes creating fair and just opportunities overall, as well as considering potential changes through an equity lens and multiple perspectives. Every community has its own culture, assets, history, and challenges on which to build. Systems and policy change efforts that do not align with a community's unique context can end up perpetuating the same inequities they are intended to correct. This includes understanding lived experiences as well as the community's culture, conflicts, tensions, and racial history.

Evaluations assess outcomes and, typically, quantitative change but oftentimes do not include qualitative measures. Qualitative measures provide context and a deeper understanding of results, such as how partnerships contributed to the community changes. Evaluations that are most helpful include both quantitative and qualitative measurements, document outcomes, and consider what works for whom and in what circumstances, as well as what facilitated the change. Mixed methods are a research approach that integrates quantitative and qualitative data collection and analysis within the same study. Mixed methods allow for a deeper understanding of how outcomes occur and provide information while changes are occurring so as to inform decision-making and mid-course corrections.

Planning how data will be used and disseminated is essential in constructing evaluations that will show impact and be useful to those creating the community change as well as to others who consider undertaking similar efforts. To best guide decision-making in real time and to also document success and impact, a mixed methods approach is well suited to collect data on the relevant constructs. Data should be collected using different modalities (e.g., surveys, interviews, observations) and from various stakeholders over time.

Components to include in an evaluation include measures of the partnership itself, tracking of systems and policy change outcomes and processes (including key milestones), resources (including in-kind contributions and new resources leveraged), stories related to change efforts, and stakeholder's perceptions of changes and impact. Tracking of systems and policy changes over time allows partners on the ground to reflect on development of change efforts. Understanding how efforts are progressing can provide insight on whether to revise strategies, approaches, or priorities. Phases of community change efforts are as follows along with what the goals for each phase:

(1) Beginning/development – Identify and agree upon goals; form strategic partnerships; ideas for change will grow in momentum in the community; base of support for the effort will grow.
(2) Proposed plan is drafted and/or introduced – A formalized plan, proposal, rule, regulation, code, or ordinance is submitted or introduced; plan is considered by the decision-making body; plan is revised or reintroduced to reflect needed changes; application for funding is submitted.
(3) Adoption – The community change effort is adopted by the decision-making body and is documented in rules, guidelines, regulations, laws, budgets, and other written documents; funding application is approved.
(4) Implementation – Funding for the change effort is appropriated by an organization, institution, system, or agency; community or school entities begin to implement the change; a process for monitoring implementation is established; capacity building includes hiring and training staff and securing space and equipment for full implementation.
(5) Maintenance/enforcement – Community change is fully implemented and efforts are underway to ensure that the change is sustained; a process for continuous monitoring and enforcement is in place; funding to sustain the change is reflected in annual budgets. There is discussion of strategies being used, obstacles that have hindered efforts, and next steps that are needed. Other community institutions and leaders must be involved to provide needed information

for assessing progress, to make mid-course corrections, and to understand how outcomes occur over time.

Outcomes of change efforts to create equitable built environments include new or improved walking paths and water trails, park improvements, and greenway and waterway development and intersections. Related local policy outcomes may include changes to existing policies or adoption of new policies or local ordinances. Examples include modifications to city, park, or transportation plans and changes to school policies. It is helpful to collect information about where these policy changes are documented (e.g., zoning regulation, city code, school policy handbook) and to include links and further information that illustrate these changes.

Quantitative outcome indicators include the number of linear miles connecting everyday destinations that were built or installed as a result of environmental or policy change and the number of new or enhanced sites or destinations connected by activity-friendly routes as a result of environmental or policy changes. Data on impact includes the number of people who have the potential to benefit from the new environmental or policy change. Although it is important to document as outcomes only those changes that have *occurred* and to not include potential or anticipated outcomes, it is helpful to document achievements and benchmarks that indicate *movement* toward these concrete outcomes. Documenting early wins builds public support for more complex policy and systems changes. Benchmarks include having strategic decision-makers/key partners on board, allocating resources, implementing action items, and in many cases uncovering new problems that must be solved in order to create the needed change.

Documentation of intermediate outcomes depicts the timeline of and progress toward longer-term changes. Successful pilot projects of policies (such as shared use agreements, Complete Streets design guides, and changes in ordinances) and built environments (such as new sidewalks, bike paths, and greenways) build credibility and capacity, allowing healthy community work to take root.

Measurement of sustainable outcomes is the ultimate goal for the evaluation. Sustainable change includes operational shifts by local government organizations that are reflected in spending priorities, new development, and built projects that maximize services and provide safe affordable access to healthy lifestyles for all residents. Sustainable change is also reflected in changes in community norms, indicated by shifts in resident demands for active living opportunities. Successful and sustainable healthy community change efforts will ultimately create a culture of health in communities where health is embedded in the community's identity.

Collecting stories associated with the work gives partnerships the opportunity to showcase the particular efforts toward change that they deem compelling. Stories convey history, identity, and culture through descriptive narrative. Stories provide a richness of detail that can inform how outcomes emerge.

Partnerships can be provided story protocols that ask for details of the specific objectives, setting, partners, resources, outcomes, impact on the community, lessons learned, and other pertinent elements that provide a qualitative picture of the process of systems and policy change. A story protocol encourages description of the partnerships' efforts including the engagement of community partners and diverse voices in the pursuit of change, an increasing ability to overcome barriers to systems and policy change, and examples of new economic endeavors associated with the partnerships' work. Tips for storytelling used in the Food & Fitness work include the following (Pyramid Communications, 2021): Every good story needs an arc – a beginning, a middle, and an end; identify obstacles that were overcome; describe any unexpected encounters or outcomes; have a hero at the center, and the hero should have a background; use quotes; keep it clear and simple (avoid jargon); and the audience should learn something from the telling. One yardstick to measure the effectiveness of a story is to consider whether the story answers questions that the audience is likely to have: Who is the story about? What do they want? What stands in their way that makes the effort interesting? How do they respond to those barriers or obstacles? What happens in the end? What does it mean?

In-depth interviews and focus groups (of 6–8 participants) allow for guided discussion/group discussion with the purpose of generating a rich understanding of perspectives, experiences, and beliefs of community members and partners involved in the work. In-depth interviews provide an opportunity to probe into the details of perspectives and information held by leaders and champions of efforts. Focus groups provide an opportunity for stakeholders to share and compare their perspectives and experiences with each other. Critical to both of these types of data collection is the selection of stakeholders and the development of questions, both of which should involve input from the community. When planning in-depth interviews and focus groups, first decide who you will talk to and what questions you will ask. Also think about how you will moderate the groups, analyze the data, and share the information obtained. The overall goal is to listen and learn.

Tracking the flow of resources can illustrate the increased acquisition and leveraging of resources in the form of matching funds, in-kind support, and additional grants and contracts to further the partnership's work over time. Documentation of resources includes the type of resource (i.e., money, full-time equivalent of work), the source, and the specific work of the partnership that the resource will support. These data enable description of the magnitude of resources and how they are deployed by the partnerships. It is helpful to verify and document the resource acquired or the commitment made (e.g., in a budget line or notation, memorandum of understanding [MOU], or a letter stating intent).

Finally, an important consideration is how information regarding processes and outcomes can be shared with stakeholders ongoingly to contribute to refinements and anticipate challenges in the work. In keeping with principles of community-based participatory research (Israel et al., 2010; Schulz et al., 1998), feedback and integration of evaluation results are critical to the ongoing work. Partnership leaders are encouraged to determine the best way to share ongoing findings with their partners and members. An example is to share findings regarding partnership engagement with working groups in order to consider which partners are not involved but may be helpful to engage in order to achieve partnership goals. Another example is to review outcomes with the partnership steering committee during the process of revising the partnership objectives. The aim is to enable partnerships to fine-tune and focus their efforts.

Dissemination of evaluation findings also informs the field of community change efforts. Evaluation has a great potential for informing others working toward sustainable changes in their community food, school food, active living, and built environment. Evaluation data that includes information about community contexts can provide both quantitative and qualitative results related to types of changes, how those changes are occurring, and how changes are both similar and unique across communities.

Efforts to capture information related to community change require that multiple stakeholders have ongoing opportunities to contribute to the documentation of the process and outcomes of the work. Best results are obtained when multiple voices are included in the evaluation and when partnership-specific work is tracked in an ongoing manner with regular opportunities for collaborative feedback and reflection. Evaluation feedback is critical to the continued development and improvement of both the work and the evaluation and valuable to others doing similar work in the community.

Case Study: REACH Lower Rouge River Water Trail

The REACH project focuses on the Lower Rouge River and three communities outside of Detroit, Michigan, Inkster, Wayne, and Westland (see Fig. 6.1), with the purpose of creating safe, equitable opportunities in these diverse communities for physical activity through blueway (water) and greenway trails. Inkster has a population of 25,366 people, with 71.6% African American and 2.9% Hispanic American. More than 33% of individuals living in Inkster live below poverty. Wayne has a population of 17,643 people, with 15.7% African American and 2.9% Hispanic American, with more than 20% of individuals living below poverty. Westland has a population of 36,506 people, with 18.2% African American and 3.5% Hispanic American, with more than 18% of individuals living below poverty (MDHHS

Chronic Disease Epidemiology Section, 2021; U.S. Census Bureau, 2019; University of Wisconsin Population Health Institute, 2021).

The partnership between the NKFM and Friends of the Rouge (FOTR) has provided a unique opportunity to align chronic disease prevention and environmental health. FOTR built momentum from strategic planning for the Lower Rouge River Water Trail, and the timing of the ongoing REACH award (2018–2023) was opportune for establishing new partnerships and staff to pursue development of the Lower Rouge Water Trail (Fig. 6.2).

The overall goal of the Water Trail is to develop a paddling network through ten cities connecting the communities via the river all the way to Detroit, Michigan. For much of the past two generations, the Rouge River has been known as one of Michigan's dirtiest rivers; it contains some of the oldest and most industrialized areas in Southeast Michigan. In 1987, the Environmental Protection Agency declared the Rouge River an area of concern under the Great Lakes Water Quality Agreement (Great Lakes Water Quality Board, 1985). This led to remediation and restoration efforts in 25 habitats that have removed pollutants, restored wetlands, increased aquatic life, and managed invasive species. Efforts of multiple partnerships, including state and local government, nonprofits, and community groups, have increased the potential of the Rouge River to provide outlets for physical activity, community connections, and a habitat for wildlife. Today, the Rouge River Gateway Greenway runs along the main branch of the Rouge River and is regularly used by pedestrians and bicycles, and families are active in its surrounding parks.

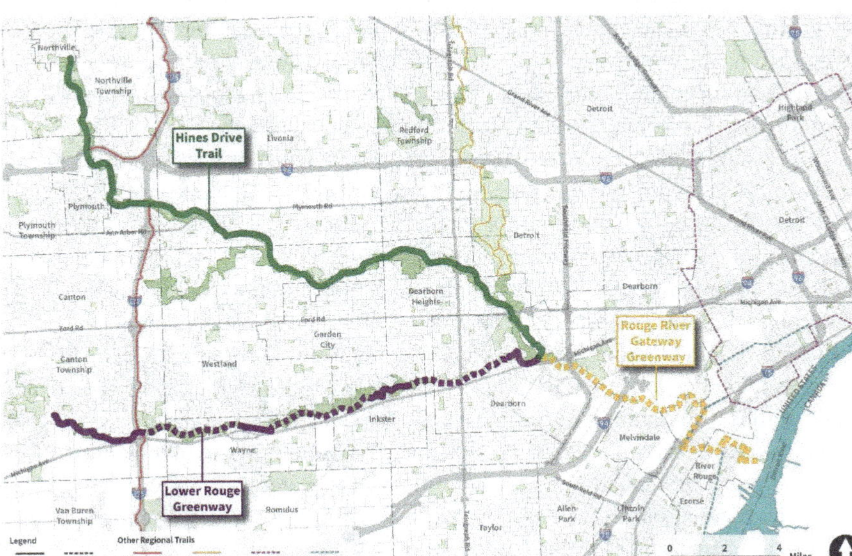

Source: Friends of the Rouge (2019). Reprinted with permission. https://therouge.org/water-trail/

Fig. 6.2 Map of the Rouge River and surrounding Michigan communities. (Source: Friends of the Rouge (2019). Reprinted with permission. https://therouge.org/water-trail/)

The Water Trail Strategic Plan, adopted by the FOTR leadership committee and published in May 2019, provided the framework for the development, maintenance, and safety of the trail, including water trail goals; alignment with National Water Trail designation, State of Michigan Water Trail designation, and Pure Michigan Water Trail designation criteria; and potential access points (e.g., from neighborhoods, parking areas, and other destinations in the community). A safety plan, drafted and published by FOTR in June 2019, covered personal safety and emergency response, including recommendations for on-the-water signage, paddling classes, woody debris management, and water quality management. Continued work will include greenway connector routes and trail amenities and building up the western part of the trail that travels through the REACH communities of Inkster, Wayne, and Westland. This case study will address this work as it relates to equity, community-driven approaches, systems thinking, assessment and prioritization of efforts, and evaluation.

The *focus on equity* in the Lower Rouge water trail development considers that poverty, racism, disinvestment in neighborhoods, and other social determinants of health create barriers for physical activity and health in the REACH communities and that improving environmental and community conditions is an important step toward changing these inequities. Many of the cities along the Rouge River, particularly Wayne and Inkster, experience financial difficulties, partially related to long-standing revenue issues, and are still pulling themselves out of the Great Recession of 2008, which decimated the tax base. Changes to the built environment in cities rely on city funding, and these financial difficulties make it unrealistic to plan on revenue from these cities for improvements. For example, a park located in a prime downtown spot in the city of Wayne has existing capabilities for a boat launch and needs significant improvement that the city cannot afford. NKFM and FOTR continue to look for grant funding opportunities for this initial investment and continue to explore equitable avenues for long-term maintenance.

The commitment to a *community-driven process* for the water trail began with the awareness that most of the initial participants engaged in the water trail development and planning were disproportionately from White, wealthier communities along the Rouge. NKFM has worked to elevate the voice of lower-income communities and communities of color in the planning and visioning process. Focus groups were conducted with members of the community to collect their thoughts about the Rouge River, and community ideas and concerns were relayed to the leadership and partners, including existing community coalitions. NKFM found ways to expand opportunities for community advocacy in the water trail planning process. An example from the city of Wayne occurred when community involvement was lacking in the public town hall planning meeting for the water trail. In response, NKFM shared the focus group information with an existing community coalition, Healthy Wayne, to garner additional feedback that was shared directly with Wayne County Parks.

Partners involved in the Lower River Rouge work along with FOTR and the NKFM include the City of Inkster through its Special Projects Division in its mayor's office. Representatives have attended Water Trail Leadership Committee

Meetings and expressed interest in strategies for woody debris management in Inkster where there are many obstructive log jams.

Along with the strategic plans that have been adopted by the REACH communities, the Lower Rouge work has uncovered the critical need to address log jams in the river and provide remediation. The Rouge River rises and falls dramatically with rain events, and this "flashy" nature impacts the river's banks and can cause trees to fall into the river and create log jams over time. Wood in the river makes a great habitat for fish and other wildlife that live in the river, but it can be a nuisance and an obstruction to paddlers. Grant funding has been targeted to help remediate and open up log jams along the planned paddling route designated by Motor City Canoe and Kayak. FOTR staff and volunteers have worked to clear log jams, and activities take place throughout the watershed to clean up debris and remove invasive species.

Woody Debris Management projects are perpetual and necessary to expand accessibility in alignment with any other type of nonmotorized trail maintenance requirements. There are opportunities for local businesses to benefit economically from woody debris maintenance. Large log jams will be opened through contracts issued by municipalities, and residents who live along the riverbanks may also want to work with companies for smaller projects that have an impact on their property. FOTR is sharing the knowledge that volunteers have acquired over the years in the management of woody debris and is exploring how to develop a certificate for businesses that go through specific training and maintain the proper insurance. FOTR is also training local foresters and landscapers who often have some of the needed equipment (e.g., chainsaws), so when log jam clearing opportunities arise, they are prepared.

The City of Westland Department of Housing and Community Development has been helpful in partnering with FOTR on the implementation of the Dry Dock Pop-Up project, which provides further opportunities for community members to contribute to the vision of the water trail and also to learn about paddling without being on the water. Recreation staff was also instrumental in helping to connect the FOTR team with Westland Farmers Market planners to help coordinate logistics and dates for a successful dry dock event, bringing the opportunity for involvement to places where community members actively congregate.

The dry dock events, piloted by FOTR in Spring of 2021, gave community members an opportunity to sit in a canoe, wear a life jacket, and gain exposure to basic paddling maneuvers without actually being on the water. The hope was to seed interest in canoeing as a form of recreation and provide information about the Lower Rouge Water Trail. NKFM has participated in these events to share information about canoeing as a way to be physically active.

Motor City Canoe and Kayak Rental, another partner, has opened livery services at Ford Field Park in the neighboring city of Dearborn (see Fig. 6.3 western section [upper map]). This has allowed FOTR to work with the City of Dearborn and Motor City Canoe and Kayak Rental as a part of the waterway network. The livery at the Ford Field Park also provides an option for beginning paddlers to practice in the pond, while more advanced paddlers can try their hand at paddling into the upstream.

Another partner, Beaumont Hospital Healthy Communities Initiative (Wayne/Westland) – Individual Healthy Communities – features formalized coalitions with membership including residents and representatives of local government, schools, business, health care, social service, and faith-based and other community organizations. Coalition members work collaboratively to optimize the reach of existing programs and support development of new programs, events, and land use that promote healthy living in their community. FOTR and the NKFM participate in the Healthy Wayne and Healthy Westland groups to promote trail use as a healthy fitness alternative and provide regular updates on trail events.

Thinking of the waterway changes as part of a larger *system,* and creating opportunities for communities to envision the complete network of the greenway and waterway development (i.e., *systems thinking*), has been critical to citizen engagement, funding opportunities, and linkages to resources at the local, regional, and state levels. The Lower Rouge River has the potential to become both a blueway and greenway with a series of community connections that can impact the health and well-being for residents in the REACH communities and beyond.

Connecting the Rouge (https://www.connectingtherouge.org/) is a regional community planning effort initiated by the Wayne County Parks that illustrates the significance of systems thinking approaches to built environment change in communities. This effort seeks to connect greenway trails to the Lower Rouge and other portions of the Rouge River waterways and greenways. This effort is critical to the identification and creation of links between neighborhoods, commercial areas, jobs, and other destinations throughout the Rouge River communities. Thinking of the complete picture of the Rouge River greenways and waterways assists in advancing the vision and core values of FOTR and supports the emphasis on their desire to develop community-led stewardship and to promote equitable access. The promotion of the Wayne County Parks and improvements to the park system also provides a platform from which FOTR, NKFM, and other partners can communicate the vision and benefits of connected trails in the REACH communities.

The *assessment/strategic planning process* involved input from all 10 communities spanning the 27 river miles. Participants included municipal leaders, NGOs, and residents. Two public meetings aimed to inform communities on the planning process, understand how participants currently use and interact with the river (e.g., where they access it, preferred activities), and solicit input on vision and future development. Feedback was also solicited on draft plans, maps, and priorities, to contribute opinions related to next steps. FOTR hosted a safety summit specifically with public safety agencies from jurisdictions across the water trail to guide education and safety recommendations. Ultimately, the Lower Rouge Water Trail Strategic Plan was published in May 2019 (https://therouge.org/water-trail/water-trail-development/). The result includes launch sites, points of interest, and design work for the entire 27 miles.

The development of the Lower Rouge River Water Trail requires coordination with other local trail projects, including the Connecting the Rouge Community planning effort that seeks to connect a regional trail network and create links

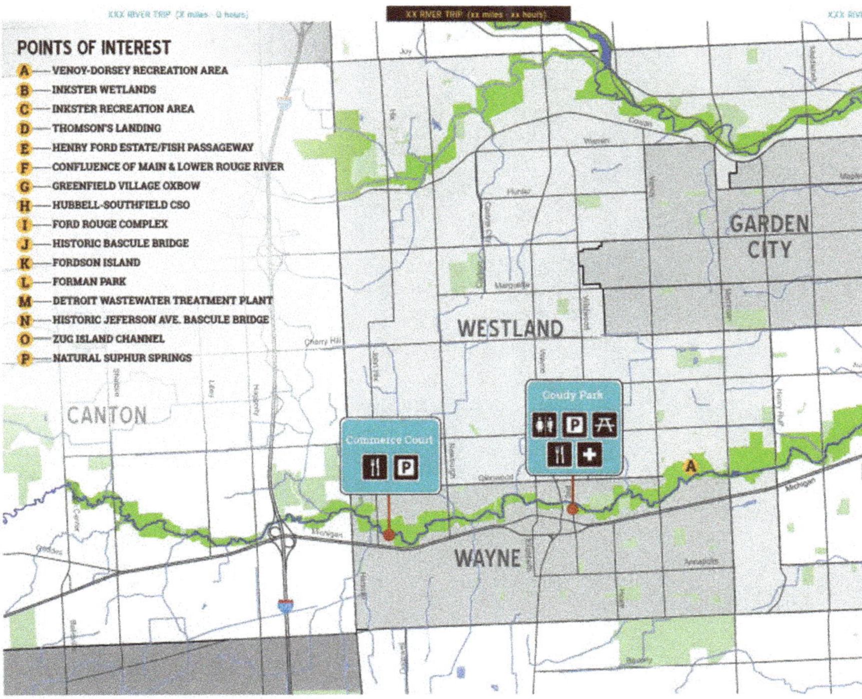

Source: Friends of the Rouge (2019). Reprinted with permission. https://therouge.org/water-trail/

Fig. 6.3 Paddling maps of the Lower Rouge River western, central, and eastern sections. (Source: Friends of the Rouge (2019). Reprinted with permission. https://therouge.org/water-trail/)

between trails and everyday destinations throughout the Rouge Valley. Facilitated by SmithGroup, Wayne County Parks engages stakeholders through the website, connectingtherouge.org, in which they can gather priorities and feedback on specific geographic features on a map. They have hosted public meetings and stakeholder meetings with cities along the river and specific nonprofit groups and have created a platform for ongoing engagement with these groups and individual community members. Their purpose is to collect ideas that will inform the continued planning and development of the Rouge River projects, with the ultimate goal of supporting vibrant, healthy neighborhoods with opportunities for economic growth.

With funding support from the Ralph C. Wilson Foundation, the Erb Foundation, and CDC REACH, the FOTR was able to hire a full-time water trail manager, who joined FOTR in February 2021. With a unique background in sports, strategic planning, creative design, and coalition building, he has brought new energy and a community-driven perspective to the development of the water trail.

Evaluation and dissemination efforts related to the REACH Lower Rouge River project are ongoing. Data on related activities are collected through tracking forms that include descriptions of the efforts, phase of the change effort, important

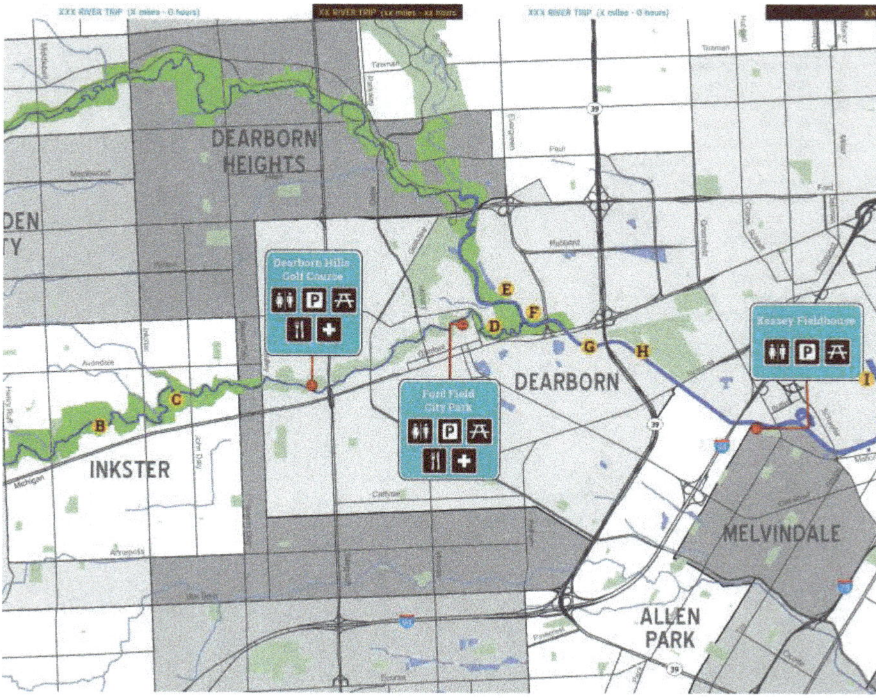

Source: Friends of the Rouge (2019). Reprinted with permission. https://therouge.org/water-trail/

Fig. 6.3 (continued)

strategies, specific obstacles that may be hindering efforts, accomplishments (benchmarks), and planned next steps. Data is also collected about partners who have a role in the efforts and who may include key community organizations, agencies, health clinics, local policymakers, businesses, and individuals/champions. Other community institutions and leaders must be involved for efforts to achieve success. The plans to gain their collaboration and the roles they might fulfill are tracked. Factors in the community or the larger environment that have helped move the effort forward are tracked. Examples include having goals aligned with individuals who are in decision-making roles (e.g., mayor's office, city council), increased public awareness about the need for change, and existing funding from multiple sources. Significant outcomes to date must be monitored including built environment outcomes (e.g., environmental changes such as new or improved walking paths, water trails, park improvements, and greenway/blueway development and intersection) and related local policy changes (e.g., changes to existing policies; adoption of new policies or local ordinances; modifications to city, park, or transportation plans; school policies). Quantitative indicators of success, such as the number of linear miles connecting everyday destinations built or installed as a result of environmental or policy change, number of new or enhanced sites or destinations connected by activity-friendly routes as a result of environmental or policy changes,

Source: Friends of the Rouge (2019). Reprinted with permission. https://therouge.org/water-trail/

Fig. 6.3 (continued)

and number of people impacted by the new environmental or policy change, are essential to evaluation efforts.

Stories are being collected throughout this work in order to document changes and further understand progress toward goals. Qualitative data will be collected in the coming year to understand stakeholder perceptions of changes in the built environment along with perceptions of how these changes have impacted, and have the potential to impact, physical activity. Data related to leveraged resources are also included in the evaluation.

Dissemination of evaluation results to partners provides information that is helpful as decisions are being made and priorities are being set. Dissemination of findings will continually be fed back to the overall partners, funders, and community throughout this work through reports, stories, and fact sheets. The intention is to make the Rouge River Water Trail a destination with the potential to spawn a blue economy through a wave, a business, and workforce development opportunities that can inherently grow from the opening of the water trail.

Summary and Implications for Family Health Research and Evaluation

The built environment, including the overall layout of the community and the quality of places, affects accessibility, use of key destinations, and overall opportunities for physical activity for families and children. The location of different activities (e.g., work, school, shopping, recreation) relative to the overall physical environment and to each other influences the fundamental character of communities and lifestyles. The plan for the built environment determines what land is developed and for what purpose, where and how far apart destinations are, what kinds of activities can happen in a given space, who can live next to whom or next to what, and what options people have for getting from place to place. Most policies related to the layout of the built environment are predicated on the use of automobiles to get around, creating daily activities and destinations that are far apart and segregated into single-use areas that are connected by major roads and parking lots. There is a strong association among opportunities for active transportation, the existence of parks/recreation facilities and programs, school environments, and level of routine physical activity and the health of children and families.

Local policy strategies can promote physical health through supportive built environments. Improvements to local parks, trails, greenways, and water ways will encourage the use of these amenities. Updating zoning ordinances, building codes, and approval processes to encourage compact community design and a tighter mixture of activities will facilitate active transportation. The resulting changes could include construction of higher density, affordable, and mixed-use projects near schools, parks, transit lines, work sites, and commercial centers. Updating road policies/standards and parking requirements/fees to improve connectivity, safety, and street design, and supporting incentives for transit and nonmotorized transportation, can contribute to supportive built environments.

Local policy change efforts are most effective when the focus is on equity; a community-driven process; systems thinking; assessment and prioritization of changes that will ensure effectiveness, efficiency, and sustainability; and evaluation and dissemination efforts that demonstrate success. In order to ensure change that is beneficial to community health, future research and evaluation related to active living and the built environment must consider all of the elements of the community change process as described here and explicitly consider efforts that will create and maintain equitable access to opportunities for physical activity for families and children. Individual health behaviors, shaped by opportunities for health in the environment, ultimately require removing obstacles to health such as poverty, discrimination, and their consequences, including lack of access to good jobs with fair pay, quality education and housing, safe environments, and health care. This includes creating opportunities for individuals to make choices in conditions that affect their own well-being.

Acknowledgments Special acknowledgment to the Friends of the Rouge, Lower Rouge River Water Trail Leadership Committee, Herman Jenkins, Sally Petrella, and Marie McCormick; Wayne County Parks; CDC DNPAO; Norwayne Community Citizens Council; and Jefferson Barns Community Vitality Center for their work on the Lower Rouge River Water Trail.

References

Acosta, M. E., Matsuzaki, M., Slater, S. J., & Sanchez-Vaznaugh, E. V. (2021). Physical activity strategies in low-resource elementary schools: Why and how are they prioritized? *Preventive Medicine Reports, 23*, 101430. https://doi.org/10.1016/j.pmedr.2021.101430

Auchincloss, A. H., Roux, A. V. D., Mujahid, M. S., Shen, M., Bertoni, A. G., & Carnethon, M. R. (2009). Neighborhood resources for physical activity and healthy foods and incidence of type 2 diabetes mellitus: The multi ethnic study of atherosclerosis. *Archives of Internal Medicine, 169*(18), 1698–1704. https://doi.org/10.1001/archinternmed.2009.302

Borrell, L. N., Kiefe, C. I., Williams, D. R., Diez-Roux, A. V., & Gordon-Larsen, P. (2006). Self-reported health, perceived racial discrimination, and skin color in African Americans in the CARDIA study. *Social Science & Medicine, 63*(6), 1415–1427. https://doi.org/10.1016/j.socscimed.2006.04.008

Braveman, P., Arkin, E., Orleans, T., Proctor, D., & Plough, A. (2017). *What is health equity? And what difference does a definition make?* Robert Wood Johnson Foundation. https://www.rwjf.org/en/library/research/2017/05/what-is-health-equity-.html

Brownson, R. C., Haire-Joshu, D., & Luke, D. A. (2006). Shaping the context of health: A review of environmental and policy approaches in the prevention of chronic diseases. *Annual Review of Public Health, 27*, 341–370. https://doi.org/10.1146/annurev.publhealth.27.021405.102137

Centers for Disease Control and Prevention (CDC). (2005). *Barriers to children walking to or from school – United States, 2004*. Barriers to Children Walking to or from School %2D%2D-United States, 2004 (cdc.gov)

Centers for Disease Control and Prevention (CDC). (2018). *Social determinants of health: Know what affects health*. https://www.cdc.gov/socialdeterminants/index.htm

Clark, N. M., Doctor, L. J., Friedman, A. R., Lachance, L. L., Houle, C. R., Geng, X., & Grisso, J. A. (2006). Community coalitions to control chronic disease: Allies against asthma as a model and case study. *Health Promotion Practice, 7*(2 Suppl), 14S–22S. https://doi.org/10.1177/1524839906287055

Clark, N. M., Brenner, J., Johnson, P., Peek, M., Spoonhunter, H., Walton, J., Dodge, J. & Nelson, B. (2011). Reducing disparities in diabetes: The alliance model for health care improvements. *Diabetes Spectrum, 24*(4), 226–230. https://diabetesjournals.org/spectrum/article/24/4/226/31855/Reducing-Disparities-in-Diabetes-The-Alliance

DeFilippis, J. (2001). The myth of social capital in community development. *Housing Policy Debate, 12*(4), 781–806. https://doi.org/10.1080/10511482.2001.9521429

Diez Roux, A. V. (2007). Neighborhoods and health: Where are we and where do we go from here? *Revue d'Épidémiologie et de Santé Publique, 55*, 13–21. https://doi.org/10.1016/j.respe.2006.12.003

Epping-Jordan, J. E., Pruitt, S. D., Bengoa, R., & Wagner, E. H. (2004). Improving the quality of health care for chronic conditions. *Quality and Safety in Health Care, 13*(4), 299–305. https://doi.org/10.1136/qhc.13.4.299

Fielding, J. E., Teutsch, S., & Breslow, L. (2010). A framework for public health in the United States. *Public Health Reviews, 32*, 174–189. https://doi.org/10.1007/BF03391597

Franco, M., Diez Roux, A. V., Glass, T. A., Caballero, B., & Brancati, F. L. (2008). Neighborhood characteristics and availability of healthy foods in Baltimore. *American Journal of Preventive Medicine, 35*(6), 561–567. https://doi.org/10.1016/j.amepre.2008.07.003

Frieden, T. R., Dietz, W., & Collins, J. (2010). Reducing childhood obesity through policy change: Acting now to prevent obesity. *Health Affairs, 29*(3), 357–363. https://doi.org/10.1377/hlthaff.2010.0039

Friends of the Rouge. (2019). *Lower Rouge River Water Trail*. https://therouge.org/water-trail/

Frohlich, K. L., & Potvin, L. (2008). Transcending the known in public health practice: The inequality paradox: The population approach and vulnerable populations. *American Journal of Public Health, 98*(2), 216–221. https://doi.org/10.2105/ajph.2007.114777

Gebbie, K., Rosenstock, L., & Hernandez, L. M. (Eds.). (2003). *Who will keep the public healthy? Educating public health professionals for the 21st century*. National Academies Press.

Great Lakes Water Quality Board. (1985). *Report on Great Lakes water quality*. International Joint Commission. https://scholar.uwindsor.ca/ijcarchive/340/

House, J. S., Lepkowski, J. M., Kinney, A. M., Mero, R. P., Kessler, R. C., & Herzog, A. R. (1994). The social stratification of aging and health. *Journal of Health and Social Behavior, 35*(3), 213–234. https://doi.org/10.2307/2137277

House, J. S., Schoeni, R. F., Kaplan, G. A., & Pollack, H. (2008). The health effects of social and economic policy: The promise and challenge for research and policy. In R. Schoeni, J. S. House, G. A. Kaplan, & H. Pollack (Eds.), *Making Americans healthier: Social and economic policy as health policy* (pp. 3–26). Russell Sage Foundation.

Institute for Democratic Renewal & Project Change Anti-Racism Initiative. (2001). *A community builder's tool kit: 15 tools for creating healthy, productive interracial/multicultural communities*. Claremont Graduate University, School of Politics and Economics. community-builders-tool-kit.pdf (wordpress.com)

Israel, B. A., Coombe, C. M., Cheezum, R. R., Schulz, A. J., McGranaghan, R. J., Lichtenstein, R., Reyes, A. G., Clement, J., & Akosua, B. (2010). Community-based participatory research: A capacity-building approach for policy advocacy aimed at eliminating health disparities. *American Journal of Public Health, 100*(11), 2094–2102. https://doi.org/10.2105/ajph.2009.170506

Keys, A. (1980). *Seven countries: A multivariate analysis of death and coronary heart disease*. Harvard University Press.

Kim, D. H. (2001). *Organizing for learning: Strategies for knowledge creating and enduring change*. Pegasus Communications.

Krieger, N. (2014). Discrimination and health inequities. *International Journal of Health Services, 44*(4), 643–710. https://doi.org/10.2190/HS.44.4.b

Kubisch, A. C., Auspos, P., Brown, P., & Dewar, T. (2010). *Voices from the field III: Lessons and challenges from two decades of community change efforts*. Aspen Institute. https://www.aspeninstitute.org/wp-content/uploads/files/content/images/rcc/VoicesfromtheFieldIII.pdf

Lachance, L., Carpenter, L., Emery, M., & Luluquisen, M. (2014a). An introduction to the Food & Fitness community partnerships and this special issue. *Community Development, 45*, 215–219. https://doi.org/10.1080/15575330.2014.901981

Lachance, L., Carpenter, L., Quinn, M., Wilkin, M. K., Green, E., Tsuchiya, K., Nelson, B., Caldwell, C., Doctor, L. J., & Clark, N. M. (2014b). Food & Community: The cross-site evaluation of the W.K. Kellogg Foundation Food & Fitness community partnerships. *Community Development, 45*, 227–239. https://doi.org/10.1080/15575330.2014.887131

Lachance, L., Quinn, M., & Kowalski-Dobson, T. (2018a). The Food & Fitness community partnerships: Results from 9 years of local systems and policy changes to increase equitable opportunities for health [supplemental material]. *Health Promotion Practice, 19*, 92S–114S. https://journals.sagepub.com/doi/suppl/10.1177/1524839918789400

Lachance, L., Quinn, M., & Kowalski-Dobson, T. (2018b). The Food & Fitness community partnerships: Results from 9 years of local systems and policy changes to increase equitable opportunities for health [supplemental material]. Appendix 5: Prioritizing potential policy & systems change strategies: Food and fitness planning tools. *Health Promotion Practice, 19*. HPP789400_Supplemental_Appendix_5.pdf (sagepub.com).

Lachance, L., Kowalski-Dobson, T., & Quinn, M. (2018c). Summarizing 9 years of the Food & Fitness initiative [supplemental material]. *Health Promotion Practice, 19*, 6S–8S. https://doi.org/10.1177/1524839918790232

Lachance, L., Quinn, M., & Kowalski-Dobson, T. (2018d). The Food & Fitness community partnerships: Results from 9 years of local systems and policy changes to increase equitable opportunities for health [supplemental material]. Appendix 6: Food & Fitness planning guide *Health Promotion Practice, 19.* HPP789400_Supplemental_Appendix_6.pdf (sagepub.com).

Landrine, H., & Corral, I. (2009). Separate and unequal: Residential segregation and black health disparities. *Ethnicity & Disease, 19*(2), 179–184. https://pubmed.ncbi.nlm.nih.gov/19537230/

Lantz, P. M., House, J. S., Lepkowski, J. M., Williams, D. R., Mero, R. P., & Chen, J. (1998). Socioeconomic factors, health behaviors, and mortality. *JAMA, 279*(21), 1703–1708. https://doi.org/10.1001/jama.279.21.1703

Larson, N. I., Story, M. T., & Nelson, M. C. (2009). Neighborhood environments: Disparities in access to healthy foods in the U.S. *American Journal of Preventive Medicine, 36*, 74–81. https://doi.org/10.1016/j.amepre.2008.09.025

Lewis, T. T., Cogburn, C. D., & Williams, D. R. (2015). Self-reported experiences of discrimination and health: Scientific advances, ongoing controversies, and emerging issues. *Annual Review of Clinical Psychology, 11*, 407–440. https://doi.org/10.1146/annurev-clinpsy-032814-112728

Livingood, W. C., Allegrante, J. P., Airhihenbuwa, C. O., Clark, N. M., Windsor, R. C., Zimmerman, M. A., & Green, L. W. (2011). Applied social and behavioral science to address complex health problems. *American Journal of Preventive Medicine, 41*(5), 525–531. https://doi.org/10.1016/j.amepre.2011.07.021

Lovasi, G. S., Hutson, M. A., Guerra, M., & Neckerman, K. M. (2009). Built environments and obesity in disadvantaged populations. *Epidemiologic Reviews, 31*, 7–20. https://doi.org/10.1093/epirev/mxp005

MDHHS Chronic Disease Epidemiology Section. Lifecourse Epidemiology & Genomics Division. Retrieved August 5, 2021 from https://www.michigan.gov/mdhhs/0,5885,7-339-71550_5104_5279-185986%2D%2D,00.html

Morenoff, J. D., & Lynch, J. W. (2004). What makes a place healthy? Neighborhood influence on racial/ethnic disparities in health over the life course. In N. B. Anderson, R. A. Bulatao, & B. Cohen (Eds.), *Critical perspectives on racial and ethnic differences in health in late life* (pp. 406–449). The National Academies Press.

National Academies of Sciences, Engineering, and Medicine. (2016). *A framework for educating health professionals to address the social determinants of health.* National Academies Press.

National Center for Chronic Disease Prevention and Health Promotion (NCCDPHP). (2020). *REACH: CDC's racial and ethnic approaches to community Health program* [factsheet]. Centers for Disease Control and Prevention. Retrieved August 23, 2021, from https://www.cdc.gov/chronicdisease/resources/publications/factsheets/reach.htm

Nieto, F. J. (1999). Cardiovascular disease and risk factor epidemiology: A look back at the epidemic of the 20th century. *American Journal of Public Health, 89*(3), 292–294. https://doi.org/10.2105/ajph.89.3.292

No Child Left Behind (NCLB) Act of 2001, Pub. L. No. 107-110, § 101, Stat. 1425. (2002). https://www.congress.gov/bill/107th-congress/house-bill/1

Papas, M. A., Alberg, A. J., Ewing, R., Helzlsouer, K. J., Gary, T. L., & Klassen, A. C. (2007). The built environment and obesity. *Epidemiologic Reviews, 29*, 129–143. https://doi.org/10.1093/epirev/mxm009

Powell, J. A. (1999). Race, poverty, and urban sprawl: Access to opportunities through regional strategies. *Forum for Social Economics, 28*(2), 1–20. https://doi.org/10.1007/BF02833980

Pyramid Communications. (2021). *Start a story strong*. Start A Story Strong | Pyramid Communications

Sacks, G., Swinburn, B., & Lawrence, M. (2009). Obesity policy action framework and analysis grids for a comprehensive policy approach to reducing obesity. *Obesity Reviews, 10*, 76–86. https://doi.org/10.1111/j.1467-789x.2008.00524.x

Schmid, T. L., Pratt, M., & Howze, E. (1995). Policy as intervention: Environmental and policy approaches to the prevention of cardiovascular disease. *American Journal of Public Health, 85*(9), 1207–1211. https://doi.org/10.2105/ajph.85.9.1207

Schulz, A. J., Israel, B. A., Selig, S. M., Bayer, I. S., & Griffin, C. B. (1998). Development and implementation of principles for community-based research in public health. In R. H. MacNair (Ed.), *Research strategies for community practice* (pp. 83–110). Haworth Press.

Schulz, A. J., Mentz, G., Lachance, L., Johnson, J., Gaines, C., & Israel, B. A. (2012). Associations between socioeconomic status and allostatic load: Effects of neighborhood poverty. *American Journal of Public Health, 102*(9), 1706–1714. https://doi.org/10.2105/AJPH.2011.300412

Schulz, A. J., Mentz, G., Lachance, L. L., Zenk, S. N., Johnson, J., Stokes, C., & Mandell, R. (2013). Do observed or perceived characteristics of the neighborhood environment mediate associations between neighborhood poverty and cumulative biological risk? *Health & Place, 24*, 147–156. https://doi.org/10.1016/j.healthplace.2013.09.005

Story, M., Kaphingst, K. M., Robinson-O'Brien, R., & Glanz, K. (2008). Creating healthy food and eating environments: Policy and environmental approaches. *Annual Review of Public Health, 29*, 253–272. https://doi.org/10.1146/annurev.publhealth.29.020907.090926

U.S. Census Bureau. (2019). *2019 American Community Survey 5-Year Estimates*. https://data.census.gov/cedsci/profile?g=0500000US26163

University of Wisconsin Population Health Institute. (2021). *County Health Rankings and Roadmaps: Michigan* https://www.countyhealthrankings.org/app/michigan/2021/rankings/outcomes/overall

Victora, C. G., Barros, F. C., & Vaughan, J. P. (2001). The impact of health interventions on inequalities: Infant and child health in Brazil. In D. Leon & G. Walt (Eds.), *Poverty, inequality, and health* (pp. 125–136). Oxford University Press.

Waters Center for Systems Thinking. (2021). *Why systems thinking?* https://waterscenterst.org/why-systems-thinking?tab=benefits

Williams, D. R. (1990). Socioeconomic differentials in health: A review and redirection. *Social Psychology Quarterly, 53*(2), 81–99. https://doi.org/10.2307/2786672

Williams, D. R. (1997). Race and health: Basic questions, emerging directions. *Annals of Epidemiology, 7*(5), 322–333. https://doi.org/10.1016/s1047-2797(97)00051-3

Williams, D. R., & Collins, C. (2001). Racial residential segregation: A fundamental cause of racial disparities in health. *Public Health Reports, 116*(5), 404–416. https://doi.org/10.1093/phr/116.5.404

Williams, D. R., & Mohammed, S. A. (2009). Discrimination and racial disparities in health: Evidence and needed research. *Journal of Behavioral Medicine, 32*(1), 20–47. https://doi.org/10.1007/s10865-008-9185-0

Woolf, S. H., & Aron, L. (Eds.). (2013). *U.S. Health in international perspective: Shorter lives, poorer health*. National Academies Press. https://www.ncbi.nlm.nih.gov/books/NBK154491/

Yu, E. (2020). Community approaches to systems change. *Journal of Public Health Management & Practice, 26*(2), 195–197. https://doi.org/10.1097/phh.0000000000001145

Zurcher, K. A., Doctor, L. J., & Imig, G. L. (2018a). Food & Fitness: Lessons learned for funders. *Health Promotion Practice, 19*(Suppl 1), 9S–14S. https://doi.org/10.1177/1524839918783974

Zurcher, K. A., Jensen, J., & Mansfield, A. (2018b). Using a systems approach to achieve impact and sustain results. *Health Promotion Practice, 19*(Suppl 1), 15S–23S. https://doi.org/10.1177/1524839918784299

Chapter 7
How Family Caregiving Negotiates and Depends on the Urban Environment

Andrew Binet and Mariana C. Arcaya

Caregiving is central to family life and crucial for healthy development, aging, and relationships, day-to-day and across generations. Caregiving is also demanding work – physical, emotional, and cognitive – the burdens of which are unevenly distributed both within households and across society in ways that reflect and reproduce social inequality along axes including gender, race, and class (Daminger, 2019; Duffy, 2011; Folbre, 2012). Since the nineteenth century, caregiving has typically been considered to take place in the private realm, a domestic activity outside the scope of public concern (Fraser, 2016). However, periodic "crises of care," such as the rapid concentration of caregiving and social reproduction in the home during the COVID-19 pandemic, have revealed just how much of this work is usually distributed more broadly throughout families' social and environmental contexts (Fraser, 2016; Stevano et al., 2021).

As a consequence of its discursive exclusion from the public realm, caregiving has remained largely absent from discussions in urban planning and design about how to spatially organize our society to promote positive social change and human flourishing. This is despite the fact that caring and being cared for are major aspects of life for most urban residents. Nevertheless, patterns of urban development since the industrial era have played a significant role in shaping the conditions under which family caregiving takes place and the possible forms caregiving can take – suburbanization being an obvious example (Hayden, 1981). Similarly, changes in the social organization of care and social reproduction more broadly, for example, deindustrialization and the corresponding movement of women into the workforce beginning in the 1970s, have had considerable influence on the development

A. Binet (✉) · M. C. Arcaya
Department of Urban Studies and Planning, Massachusetts Institute of Technology, Cambridge, MA, USA
e-mail: abinet@mit.edu; marcaya@mit.edu

S. E. Ortiz et al. (eds.), *Environmental Impacts on Families*, National Symposium on Family Issues 12, https://doi.org/10.1007/978-3-031-22649-6_7

trajectories of cities (Curran, 2017; Parker, 2017; Winant, 2021). The reciprocal relationship between caregiving and the urban environment suggests that it is important for urban planners and designers, as well as those in allied fields like architecture or public policy, to develop frameworks for attending to and engaging in matters of care. Doing so is crucial for ensuring that urban environments promote and contribute to a plurality of healthy familial relationships, community resilience in the face of challenges like housing insecurity and climate change, and health and social justice more broadly.

Care is "a species of activity that includes everything we do to maintain, continue and repair our world so that we can live in it as well as possible. That world includes our bodies, ourselves and our environment" (Tronto, 1993, p. 103). The breadth of this definition of care reflects the fact that care is ubiquitous and a "defining condition of what it means to be human" (Kleinman, 2015, p. 240). It is a vital activity engaged in the everyday maintenance of the web of life, enacted through specific relations of interdependence, and grounded in moral commitments. Care is necessarily relational in that it involves responsibility and commitment to an object of care. This relation is not typically purely dyadic but instead involves a complex network of actors, resources, and actions, including elements of the environment where care takes place. Because care is relational, geographers have argued that care is spatialized and that care settings – including private spaces like the home, professional spaces like clinics, and public spaces like parks – form part of the social fabric of the places where they are located (Conradson, 2003; Lawson, 2007; Milligan & Wiles, 2010). Care can thus be understood as a quality of spaces that is produced relationally (Bondi, 2003).

Today, societies around the world are faced with a "crisis of care" (Fraser, 2016). States are increasingly "externalizing care work onto communities and families while simultaneously diminishing their capacity to perform it" through systematic disinvestments in social welfare policies, programs, and institutions (Fraser, 2016, p. 112). As a result, care is increasingly commodified for those who can afford it and an individualized, privatized responsibility for those who cannot. These patterns are global, but in the USA, they are accentuated by an aging population, uneven and ungenerous welfare state provisions, structural racism, and other contextual factors (Black, 2020; Poo, 2016; Roberts, 1997). Every form of capitalist society harbors social reproductive contradictions, Fraser argues, because social reproduction is the "condition of possibility" for capital accumulation. Capitalism's orientation toward accumulation, though, tends to destabilize and deplete the social reproductive processes on which it relies (Fraser, 2016; Mohandesi & Teitelman, 2017). Under neoliberalism, "endemic care deficits" have become normalized, and forms of care that do not serve the interests of value extraction and profit are actively undermined as individuals are pushed to take more and more personal responsibility for their own well-being under austerity (Tarrant, 2018; The Care Collective, 2020, p. 9).

The contemporary crisis of care is also both reflected in and exacerbated by the built environment. After all, planning responds to capitalism's crisis tendencies

(Lefebvre, 1996). We can see this in the growing care needs of an aging population, many of whom are spatially isolated from family and friends who might feasibly be able to care for them, and in urban planning's push to develop new means of "aging in place" (Biglieri 2017; Forsyth et al. 2019). We can see it also in the fact that workers in the care sector, the fastest-growing part of the labor market, increasingly cannot afford to live in major cities or must work more than one job to be able to do so, which in turn puts the quality of care they give in jeopardy. Further, the privatization of public spaces is eroding capacities for community care (The Care Collective, 2020). From an urban planning perspective, alleviating the rising pressures of the "crisis of care" might thus involve coordinated policies that seek to increase the value of paid care work. Simultaneously, urban planners and policymakers should invest in infrastructures that shift the burdens of care and social reproduction away from individual households and back onto the shoulders of the collective, and the city itself, in ways that reduce the time and resources people must devote to unpaid care work and the cognitive labor that undergirds this work (Binet et al., 2022; Daminger, 2019; Duffy et al., 2013; Miraftab, 2021; Randles, 2020).

A crucial first step in understanding how to intentionally plan for and develop urban environments that foster and support more equitable relations of care at the societal level, and in reducing the difficulty and burden of care at the household level, is developing a more systematic understanding of how the work of everyday caregiving engages with and is shaped by the urban environment. With a deeper understanding of the aspects of care work that rely on the urban environment, it becomes possible to assess how historical or planned changes in urban environments affect these domains of care work: Do the changes make care work easier or harder, more or less resource-intensive, more individualized or collectivized? Do the changes reduce or expand caregiver's choice and agency? Similarly, such an understanding stands to help clarify planning, design, and policy priorities for urban development that makes neighborhoods and cities more livable for families and promotes family well-being.

This chapter draws on interviews conducted with caregivers in the Boston, MA metropolitan area to explore how the day-to-day work of care interacts with, responds to, negotiates, depends on, and is shaped by different social and spatial aspects of the urban environment. It emerges from a broader project exploring how matters of care mediate the relationship between health and place in cities (Binet, 2021). That project was conducted as part of the Healthy Neighborhoods Study (HNS), a longitudinal participatory action research study of the relationship between neighborhood development and community health in nine Boston-area neighborhoods. In the next section, we describe the HNS study design and our research methods, followed by the results of our analysis. We conclude by urging urban planners and policymakers to account for care in the design and evaluation of urban interventions and to prioritize making the work of care easier, to make cities more livable for families and more conducive to family well-being.

Methods: The Healthy Neighborhoods Study

We explore how family caregiving depends on and negotiates the affordances and shortcomings of urban environments through an analysis of qualitative data collected from semi-structured interviews with nearly 70 caregivers. These interviews were conducted as part of the Healthy Neighborhoods Study (HNS), a longitudinal multisite participatory action research (PAR) project that began in 2016. The HNS explores the relationship between urban development and community health in nine gentrifying neighborhoods across the Boston metropolitan area (Arcaya et al., 2018).

PAR is an approach to inquiry that centers the lived experiences of those individuals and communities most affected by the phenomena under investigation across each phase of research and seeks to use the knowledge cocreated through research to act on and improve the community's social and material conditions and the power relations that shape them (Greenwood & Levin, 2006). The HNS PAR model is centered on a team of about 45 Resident Researchers who work in collaboration with a local partner organization in each community (Fig. 7.1; The Healthy Neighborhoods Study, 2020). Resident Researchers are community residents recruited by community partner organizations and participate in all aspects of the research process, including research question development, instrument design, data collection, data analysis, sharing research findings with various audiences, and using study findings to develop local interventions addressing the health impacts of neighborhood development (Binet et al., 2019). Resident Researchers range widely in age, education, familiarity with research, and life experiences. Resident Researchers are trained in PAR methods and research ethics and are registered as IRB (Institutional Review Board)-approved study personnel.

During the first year of the HNS, we began to explore the relationship between place and health through workshops with Resident Researchers and site coordinators from community partner organizations. We collaboratively developed a set of three overarching research questions to guide us through our research-to-action cycle:

1. What changes in experiences, opportunities, health, and quality of life are residents experiencing as their neighborhoods change?
2. How does what is built, how it is built, and for whom it is built impact health?
3. What roles do residential mobility and displacement play in changes in population health?

While care and social reproduction are not an explicit focus of the HNS, the themes pervade the study. The impacts of neighborhood changes on parents, children, families, and kin are recurring themes in conversations and workshops with Resident Researchers, many of whom are themselves caregivers whose own experience of gentrification has been shaped by this role. Indeed, Resident Researchers' involvement in the HNS may be seen as a form of care for the well-being of their neighborhood and community: many discuss their role in terms of meeting important community needs and helping to heal collective wounds. In collaborative research design and data analysis workshops, our team's discussions of changes to the built

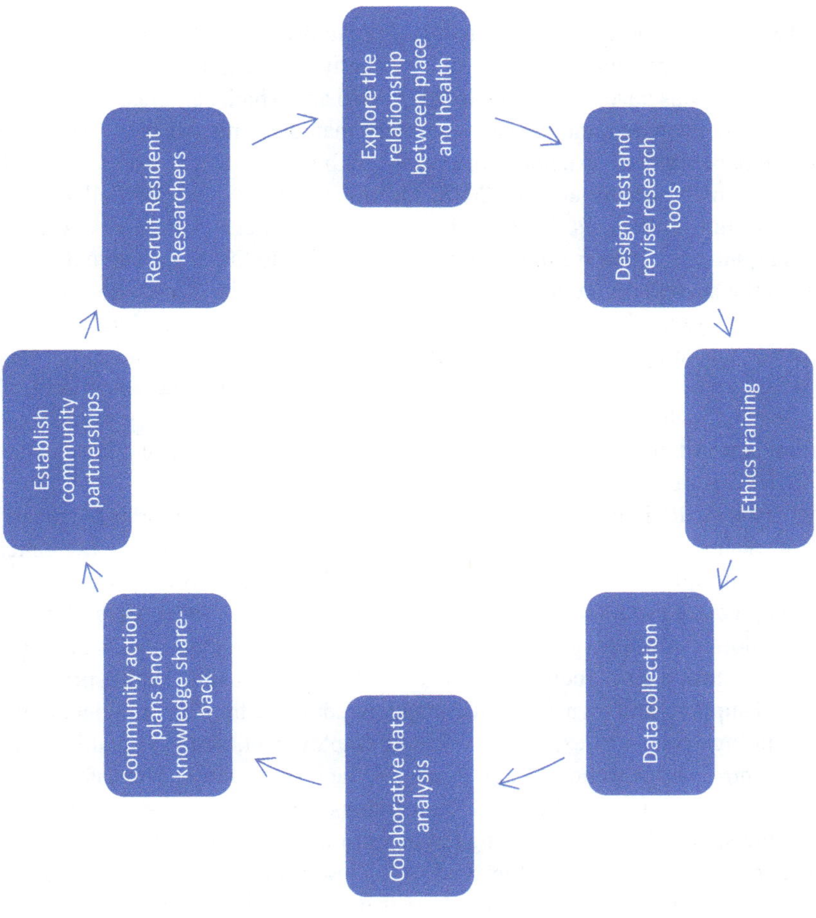

Fig. 7.1 The HNS PAR process cycle

environment in domains like housing and green space are often explicitly focused on how these changes impact care work. The same is true of our conversations about factors that might mediate the relationship between gentrification and health, such as financial security, social support, and one's ability to meet their priorities in life.

Data Collection

The HNS collects both quantitative and qualitative primary data from respondents living in the study communities. We collect quantitative data through a community survey tool that was developed and piloted in 2016 and which has since been fielded three more times on an annual basis with more than 3000 respondents. We collect qualitative data via semi-structured interviews with members of a nested longitudinal cohort which we initiated in 2018. During the first round of collaborative research design workshops in 2016, Resident Researchers expressed interest in gathering qualitative data from respondents in addition to the quantitative data collected via the survey. As a consortium, we agreed to prioritize survey development during the first year of the study and then develop an interview tool during the second year of the study based on our experience fielding the survey and initial findings. Here, we focus on the process for designing qualitative data collection tools and collecting qualitative data; descriptions of processes for collecting primary and secondary quantitative data are described elsewhere (Arcaya et al., 2018; Binet et al., 2019; Daepp et al., 2021).

The "Round 1" interview tool was developed during collaborative research design workshops held in Fall 2017. During these workshops, Resident Researchers discussed how qualitative and quantitative data differ, and how qualitative data could complement the quantitative data they gathered via the survey. They identified six survey domains where qualitative inquiry would advance their understanding of the study's central themes: neighborhood and belonging, housing, health, prioritization, social support, and ownership of change. In addition, they identified a number of crosscutting themes to explore: work and employment, sense of stability, and sense of control and direction in life. Academic facilitators drafted interview questions based on input from Resident Researchers about which aspects of these domains we should focus on learning about from interview respondents. Resident Researchers gave feedback on the draft interview protocol, and some Resident Researchers also agreed to be test interviewees in order to assess how well the questions worked. Final revisions were made according to their feedback. While two subsequent rounds of interviews with cohort members have been conducted, designed following the processes similar to that described above, here we analyze data from Round 1 interviews only.

In early 2018, the HNS began operating a nested longitudinal cohort of 150 people. Cohort members were enrolled in the study in one of two ways. First, cohort members could be enrolled by Resident Researchers during the course of regular surveying: Resident Researchers would offer every third respondent the choice of joining the longitudinal cohort and receiving an interview and survey at a later date or completing the anonymous community survey. Alternatively, cohort members could be enrolled on "rapid enrollment" days when study personnel would station themselves at a community hub (e.g., a library or YMCA), identified by Resident Researchers and community partners in that neighborhood, and recruit as many cohort members as possible. If individuals agreed to be enrolled in the cohort, their contact information was collected, and study personnel would reach out to them in the coming weeks to schedule an appointment for an interview and survey at a convenient time and place of their choosing, usually either their home or a nearby café or library. In some cases, the survey was administered when enrollees' contact information was collected. If interviewees consented to be recorded, interviewers did so using either an application on their mobile phone or via stand-alone recording device. Audio files of interviews are professionally transcribed.

Analytical Subsample

The analysis we report on here is based on a subsample of cohort members who play caregiving roles in their everyday lives. The sample was identified using the only indication of caregiving responsibility on the HNS survey: whether a respondent has children. The sample is not restricted to biological parents and includes grandparent caregivers. Interview responses show that some members of the sample also care for other people in addition to children like elderly relatives or neighbors. Descriptive statistics about the analytical subsample are presented in Table 7.1. Three quarters of the respondents are women. Nearly 80% of the respondents are people of color, and 50% are black. Most respondents report that their health is "good" or better, but one in five say their health is only "fair" or "poor." More than half of respondents have an income of less than $30,000 per year. The sample is thus reflective of who bears the greatest burden of care work in the USA: mostly women, mostly people of color, and mostly poor and working-class people.

Table 7.1 Analytical subsample descriptive statistics

Variable	Count	Percentage (%)
Gender		
Male	16	24
Female	51	75
NA	1	1
Race		
Black	33	49
Latinx/Hispanic	8	12
Multiracial	8	12
Others	4	6

(continued)

Table 7.1 (continued)

Variable	Count	Percentage (%)
White	14	21
NA	1	1
Age		
18–24	1	1
25–34	16	24
35–44	17	25
45–54	10	15
55-64	15	22
65+	7	10
NA	2	3
Self-rated health		
Poor	1	1
Fair	13	19
Good	30	44
Very good	16	24
Excellent	6	9
NA	2	3
Annual income		
Less than $15,000	19	28
$15,00–29,999	17	25
$30,00–59,999	16	24
$60,00–89,999	6	9
$90,000–119,999	2	3
$120,000 or more	0	0
Do not know	3	4
NA	5	7
Neighborhood		
Brockton	5	7
Chelsea	8	12
Dorchester	13	19
Everett	3	4
Fall River	7	10
Lynn	1	1
Mattapan	4	6
New Bedford	11	16
Roxbury	16	24

Analytical Strategy

The analysis conducted for this chapter is based on three rounds of qualitative coding. First, we used open coding on a subset of ten transcripts to generate a preliminary understanding of key themes in the data and began to write analytical memos about these themes as they developed. Once we believed that we had a grasp of the main themes that were emerging, we developed an analytical codebook based on a synthesis of the open codes we had generated so far. For the second phase of the analysis, we used this codebook to recode the transcripts we open-coded, as well as to code the remainder of the transcripts in the subsample to the point of thematic saturation. During this round, we further developed our analytical memos and began to speculate about the relationships between different themes and the emerging theoretical structure of our results. In the third phase of the analysis, we conducted axial coding within specific significant themes in the analysis to better explore their respective dimensions.

In the analysis, we sought to look at respondent's lives from the perspective of care. We analyzed each respondent's transcript from the perspective of their role as a caregiver, inasmuch as we might analyze their transcript from the perspective of their residency in a particular neighborhood. We did so in order to attune ourselves to the ways that caregiving shaped respondents' perceptions across the range of interview themes, from financial security to neighborhood belonging to priority fulfillment.

While the large majority of respondents in the analytic subsample are caring for their own children, in reporting the results, we have used the labels of "caregiver" and "dependent" rather than, for example, "parent" and "child." Not every caregiver in the sample is a parent; for example, some are grandparents. Similarly, not every caregiver is caring only for children; for example, some are also caring for siblings or elders. Furthermore, the focus is on care as a type of work that stretches across different types of relationships of interdependence, rather than on parenting or children per se.

Results: Caregiving Strategies Depend on the Urban Environment

We report on one key dimension of the analysis: patterns of caregiving strategies that depend on and/or negotiate different characteristics of the urban environment that caregivers employ to fulfill important caregiving goals for their dependents, for their communities, and for themselves.

Caregiving strategies refer to the planned or coordinated actions respondents undertake in an effort to maintain the conditions and efficacy of their care and to accomplish their caregiving goals. Our analysis focused on strategies that engage environmental features and qualities in caregiving decisions and practices and/or which seek to change some dimension of the environmental context of the caregiving relationship, in service of meeting a respondent's caregiving goals. Our analysis identified six patterns of environmentally-dependent caregiving strategies: managing the dependent's relationship to their environment, securing and maintaining the material conditions of care, building and maintaining relationships that support caregiving, preserving dependent and caregiver health, time management, and balancing work and caregiving. Together, these strategies show how the work of caregiving is shaped by material and social characteristics of the urban environment that determine the level of control that caregivers have, the choices they can or must make, and the aspects of caregiving that require resources, time, and attention.

Caregiving strategies were shaped by numerous environmental, social, and economic constraints in addition to inequitably distributed opportunities that reflect structural racial and class inequities at the societal level, as well as the context-specific stratification of social reproduction. As a consequence, not all caregiving strategies can be understood as freely chosen. In the strategies they undertook, caregivers were often trying to overcome the shortcomings of their context, doing so by balancing between what they deemed to be most effective given their circumstances and their knowledge from past experience. One respondent explained it thus: "What's important for me and my family at the beginning of the day is always at the top of my list. How do I make that better? I minimize what I don't need… What I try to do at the most is get what's important done at that particular time. And I work towards getting that one thing done, and then work on the next thing."

Moreover, not all caregiving strategies were undertaken by the caregiver alone. Many relied on various forms of social and institutional support. However, typically, caregivers were nevertheless doing the work of securing and marshaling these sources of support to carry out strategies in service of their goals. Access to these sources of support varies in numerous ways, including according to tenure in one's community, access to information, languages spoken, immigration status, racism and discrimination, and the energy and time available to maintain relationships with sources of support. Given that one of the strategies is creating support systems, the importance of social support for carrying out caregiving strategies also illustrates how strategies can be dependent on one another.

Overall, these caregiving strategies relate to the affordances of the urban environment in three ways. First, caregiving strategies depend and rely on relevant affordances of the urban environment when they are accessible. Second, caregiving strategies involve the work of coordinating across different elements of the urban environment to establish connections and synergies between components that may not be built, designed or planned but which are necessary for the caregiver to meet their caregiving goals. And third, the strategies entail creating, replacing, or compensating for those aspects of the urban environment that are not accessible or adequate. Thus, the amount of work that these strategies actually involve, and the difficulty of that work, is shaped by the underlying characteristics of the urban environment. Next, we walk through each pattern of environmentally dependent caregiving strategies in turn.

Managing Dependents's Relationship to Their Environment

The first pattern of caregiving strategies in our data concerned caregivers' efforts to manage and modulate their dependent's relationship to their neighborhood environment. Such strategies included efforts to cultivate the dependent's exposure to elements of the urban environment that the caregiver deemed beneficial, or to protect the dependent from aspects of the urban environment deemed detrimental to well-being and development. For example, one caregiver described protecting her dependents from perceived threats to their well-being from their neighborhood social environment. She is concerned about "drama" around the neighborhood such as violence and public drug use "that could potentially harm my kids somehow." She said her children like to play outside, but in her neighborhood, "There's nothing for kids to do…If there's a park or something, there's always some drama happening there. If they ever open something, they usually close it. So, kids really have nothing to do. So, I feel like [the neighborhood] should just have something…where we can bring our kids. That's how you make friends…. But there's nothing to do." She explained that as a result, her family mostly stays inside the home because "they're little and I want to keep them as far as I can from everything that's happening until we move out of here and find better." Unfortunately, protective dispositions toward perceived threats to dependent well-being from the neighborhood environment often resulted in increased isolation and withdrawal for both caregivers and their dependents.

Strategies for managing the dependent's relationship to their environment also included efforts to build dependents' understanding of how their neighborhood, city, and society work, with a view toward cultivating the dependent's ability to navigate these contexts with some degree of independence. A caregiver in a different neighborhood explained that she intentionally takes a more exposure-oriented approach to managing her dependents' relationship to their environment. "They're already exposed anyway, so I expose them to everything." For caregivers in this study, exposure could take the form of involvement in various aspects of the community, whether in volunteering or activities. Exposure could also take the form of interpreting the environment and educating the dependent about what they experience. "I get my kids involved in a lot of activities. They do…basketball, soccer, street hockey, anything. I just get them involved…," the caregiver elaborated. She described how she brings her kids along when she volunteers in the community and expressed a desire for more organized volunteering opportunities for children.

Similarly, one respondent shared an anecdote from a walk with her son in their neighborhood. "We were walking to KFC, and a gentleman there, his shoes were so ran out, and my son says, 'Mommy, why's his shoes look like that?' And I said, 'I don't know'. I said, 'Don't ever knock a homeless person because you never know why they became like this. They could have probably had an executive job, working on Wall Street somewhere, and shit just happened.' And I said, 'Everybody has a story. We just don't know his.'" Awareness was a key priority for caregivers, and spatial context and public life necessarily became the means through which this awareness was cultivated.

For some caregivers, involving dependents in the neighborhood could help fulfill the goal of protecting them from some aspects of the environment. One respondent explained, "Unfortunately, [my son] can't hang out in his community. That's the downfall of our community. My son can't go out and hang out on the block. I could in the projects," she said, reflecting on her own upbringing in public housing in the same neighborhood that has since been demolished. "I could go hang out from sunup to sundown in the projects because there was always a parent looking out for you. Unfortunately…today, not everybody knows who's everybody's children…. He's in something organized at all times because I don't want him diverting. And I know this is not a guarantee that he won't divert, but at least he knows he has options."

Another caregiver in the same dense urban neighborhood described the importance of safe public spaces for ensuring that dependents have positive exposures to their neighborhood environment. "They're not able to just go outside and play freely, like if you had a house and a backyard with a fence…so a lot of them stay inside. So, you have to take them. We have a lot of amazing parks. Actually, yesterday, we took [my daughter] to seven different parks." Often, however, protecting dependents from environmental influences leads to staying home, when possible, which typically leads to social isolation for the dependent, caregiver, or both. "You want to get in, get out, and be careful where you're going," said one caregiver.

Securing and Maintaining the Material Conditions for Care

A second major group of caregiving strategies were respondents' efforts to ensure the availability and suitability of basic material necessities for caregiving, including shelter, food, diapers, and clothing. This group of strategies depended the most on economic and resource aspects of the urban environment. These strategies required caregivers to navigate both the affordances (or lack thereof) of the physical environment and structural limits on the availability of goods necessary for care, such as a lack of affordable housing. Often, caregivers' efforts to secure the material conditions of care combined market-based strategies with efforts to find free or subsidized goods through community resources and programs, which required flexibility. As one caregiver explained, her strategy for securing basic goods "depends a lot on cost. We take free opportunities and change plans."

Many respondents were engaged in efforts to improve their housing situation, which usually involved navigating both the private rental market and long waiting lists for housing subsidies or public housing. Overall, housing was an area where caregivers often had very little choice or control over their circumstances, and many had to strategize to build the rest of their lives around the housing options that were available to them. One caregiver's housing story exemplifies this pattern. When asked why he and his family lived where they did, he said, "It's pretty much I didn't have a choice. I was kind of forced out here." He explained that he and his family were living in a gateway city north of Boston and that his girlfriend lost her job. "I couldn't afford the rent so we basically had to move... The only place we could find at the time, the rent was basically too high...but it was either that or go into a shelter...So we tried to deal with it, and we was there for about a year and just fell off with the rent, couldn't do it, and basically ended up in a shelter before we moved out here." Staff at the shelter helped him find his family's current apartment. Multiple respondents reported benefitting from programs that helped them find housing when they were living in a shelter and from programs that provided short-term rental assistance and basic furniture for parents expecting newborns.

Providing food for dependents was another important strategy for ensuring the basic material conditions for care and often involved combining different types of neighborhood resources to make ends meet. One respondent described how she combined different food procurement options in her neighborhood to meet her family's needs: "You've got to have the right money to buy the right food to keep you healthy, and the stuff in the store is so expensive, [and] I haven't had food stamps in over 20 years. What I do, I try to catch the $2 food bag truck...get me two or three bags, and sometimes I talk to the social workers in the hospital and they're giving me a gift card or something. But it's really hard because the stuff that's healthier for you is more expensive, and the stuff that's bad for you is more cheaper in the store."

Respondents also navigated community-based resources to get other basic goods for free. For example, one respondent said, "I know every single food pantry out here possible. I know everywhere to get deodorant, shampoo, toothpaste. I know where to go to get everything...your basic body needs. I know where to go to get your little house essentials. I know where to go to get diapers, wipes... So, I go all around, and I stock up on everything...."

Creating Support Systems

Social support was a crucial resource that many caregivers relied on to fulfill their caregiving goals and responsibilities. The strategies they used to build and maintain social support systems depended on physical aspects of the urban environment like accessibility and proximity, as well as social factors like relationships with neighbors. These strategies were shaped by the availability or absence of different potential sources of support in the respondent's local context, since the support required was typically hands-on in nature, like child care, transportation, or food preparation. These systems of support were described by respondents as being crucial, both to the continuity and quality of their dependents' care and for the well-being of caregivers themselves. Some respondents also connected these efforts to the task of building community-level social cohesion more broadly. Others remarked upon the ways that the absence of social cohesion at the community level inhibited their ability to form trusting relationships of social support, particularly in cases where respondents were relatively new to their neighborhood or did not grow up there.

Respondents often sought to enlist other family members in their caregiving support systems. For example, one caregiver described how she trained her family members who live nearby in the skills necessary to help care for her physically disabled daughter. "That's very helpful when they come and ask to keep my daughter while I can go do a day with my son one-on-one. That is all a blessing. And that's a lot of support." She also described how family members would bring over food and help with tasks like cooking and laundry and provide services like hairstyling that the respondent would otherwise have to pay for. This is an example of how physically proximate support systems can benefit both the dependent and the caregiver. Similarly, another caregiver counted on her children's grandparents, who lived close by, for childcare support when her and her partner's work schedules overlapped. A number of respondents reported strategizing about housing decisions based on family-oriented social support systems that they wanted to have in place. One respondent described waiting to move until she and her son no longer needed to rely on his grandmother for day-to-day childcare support, and another explained that she chose to live in a community where she does not have many friends and feels isolated, because her partner's mother was able to provide them with housing and assistance with childcare.

Respondents also reported trying to build supportive relationships with neighbors. One caregiver described how she and some of her neighbors take turns organizing activities and outings for their dependents. "We manage things for them to be happy, for having friends because it's really important when you're in a neighborhood to have friends around." Likewise, another caregiver explained, "My neighbor, she… just went to get her kid at school and I'm still with [her] baby for her. And the neighbor over here, next to us, my kids were…outside picking up all the leaves and they decided to do her front yard too. So we help each other." These relationships may also come in handy in a crisis, as in the case of one caregiver who told how she had to organize neighbors in her building to sign a petition on her behalf when she and her son were at risk of losing their housing.

Some caregivers were also involved in creating support systems at the community level. Examples include one respondent's efforts to set up a support group for single parents and another's efforts to provide employment and support for people who had experienced sexual harassment. Such efforts could also be more ad hoc in nature. For example, one caregiver shared how at the time of the interview she was sheltering a homeless family while they got back on their feet. These examples demonstrate the myriad ways that some caregivers seek to build forms of mutual dependence that rise to meet one another's needs for support in the absence of coordinated support for caregivers from institutions and policies.

Unfortunately, many caregivers felt like they did not benefit equally from the support systems that they helped to create and maintain. The caregiver who was sheltering a homeless family explained, "I help a lot of people. But in turn, I don't feel like there's nobody there for me and that's probably the worst place to be in. Especially when you get older and you know that you're getting a little bit more tired and you just wish that you had a little bit more support in certain areas. And I just don't really feel like I have that…especially on a level where like, the level of positive things that I really want to be around and surround myself around. And it's just not always there... It makes me sad." The experience of giving more social support to family and friends than they were receiving was extremely common among respondents.

Finally, some caregivers also strategized to secure more formal sources of support for their dependents and themselves, like therapists, social workers, faith-based support groups, and parenting classes.

Preserving Health

Unsurprisingly, respondents described a range of caregiving strategies related to preserving health – both their dependents' and their own – which were shaped by the qualities and affordances of their urban environment. Health-oriented strategies related to the urban environment typically sought to protect against or ameliorate health threats that caregivers perceived to be stemming from their environment. Other health-oriented strategies sought to cultivate and maintain healthy behaviors by taking advantage of the environment. Some of these strategies, such as accessing healthy food, overlapped with those described earlier as relating to securing the material conditions for care.

Some respondents were focused on addressing health concerns in their housing. One respondent explained that she was struggling with housing-related health challenges such as mold and rodents. She was cleaning constantly to ensure that her grandchildren, for whom she is a full-time caregiver, are safe when they play on the floor of her apartment. She said both of her grandchildren have asthma and allergies and that she thinks these problems are related to the condition of their apartment building. Two respondents reported that their dependents had elevated lead levels in their blood from previous housing. One described how she was trying to save money to move to a new apartment but at the time of the interview had not yet been able to secure new housing. Another described how, after living in a shelter, she and her children were placed in an apartment with lead paint, plumbing issues, and mold, and she had to take her case up with the city in order to be transferred.

Respondents also worked hard to preserve their own health, efforts that were typically described as being for the benefit of their dependents as much as for the respondent themselves. One caregiver explained that because she was now raising her grandchildren, she was trying her best to take good care of her health by walking around the neighborhood, going to classes like yoga, and keeping on top of her medical care. Her actions demonstrate how much caregivers' efforts to maintain their health depend on access to salutogenic neighborhood resources and amenities. Another respondent showed how caregivers also rely on the urban environment to maintain their mental health. He described taking long walks through Boston to visit different churches – which he called "pockets in the city" – when he was feeling stressed or overwhelmed about his new caregiving responsibilities, having recently taken custody of his adolescent daughter. "I'm a prayer person. I like to meditate. So, I think that's a priority for me. If I plan on pulling this whole [parenting] piece together, I think that spirituality has to be the foundation that gives me that guidance to go forward in this journey," he said.

Sometimes, caregivers' efforts to preserve their own health were oriented toward protecting dependents from stress about their caregivers' well-being. For example, one respondent shared that because of a health condition that impairs her mobility, "I have limitations that I didn't have before…So, I have to gauge and monitor how much I can do, because I have a [teenaged] son at home who lost [his] father. I'm his sole surviving parent, I don't want him to have the stress of concern of whether mom's okay." Another caregiver shared a similar sentiment when describing her effort to hide the financial challenge of affording healthy food from her son. "I think that what we know about child development, especially, is that if the child is worried about their basic needs or if they're seeing their parents or caregiver worry about their basic needs, they don't develop properly."

On the other hand, sometimes respondents had to sacrifice their own health for the benefit of their dependents. For example, one respondent explained that the demands of caring for her dependents meant that "I'm probably the least active out of everybody, but I'm active in different ways because I'm always cleaning. I'm always doing things and working on the weekend…So it's like you got to keep yourself moving… I have to take more focus for me in order to do that. But I don't do that because I care too much for eeny, meeny, miny, moe and all the birds and the animals…".

Time Management

Caregivers reported being engaged in various forms of spatiotemporal choreography to ensure that there were no gaps in the day-to-day continuity of care, which was of special concern for those caring for younger children and people with serious health needs. Successfully figuring out the logistics of demands on caregivers' time was usually a relief. One caregiver spoke to the importance of coordinating different aspects of her responsibilities as a caregiver across time. "I feel much more confident now that I have this [new job], even though it's not the line of work that I expected to be in or really would like to be in. It works for me. The hours are the exact same as my son's so I don't pay for a babysitter. I don't have to worry about daycare or whatever. So yeah. It's helped."

Some time management strategies were day-to-day. Respondents reported having schedules and routines to help manage childcare, activities, school, meals, and the urban mobility necessary to accomplish these activities within the space of a day. Respondents relied heavily on routines in order to fulfill caregiving responsibilities, which in turn depended on reliable accessibility and mobility in the city. "We're very regimented, I guess, as far as that goes, in order to keep us on track, emotionally healthy and physically," one caregiver explained. When routines were thrown off, some respondents turned to their support networks for assistance, and those who did not have access to support faced consequences, like lost income, when they needed to prioritize caregiving over work as a result. For example, one respondent described the routine of sharing childcare responsibilities with her husband; one of them worked an early morning shift while the other worked in the afternoons and evenings to ensure that they are available for childcare as much as possible. Fortunately, when their routine broke down, her mother-in-law who lived nearby could provide childcare support in a pinch.

Other time management strategies that caregivers employed were longer-term. Often, these involved keeping things afloat while waiting for circumstances to change, especially housing circumstances. One caregiver explained, "I'm just trying to find a good job and better housing. I'm hoping that somebody reaches out to me soon because I'm on a lot of [waitlists]...So I'm just praying that within 5 to 10 years, I have better housing..." Another caregiver said she had been on a waitlist for about a decade for a different apartment that could better accommodate her daughter's disability. "They said you've got to be on it for 10 years in order for something to actually come through for you because there's so many people. Because every time you get close and closer, you turn on the TV and someone's been burnt out [in a fire]; someone's homeless, so they come first because of the simple fact that you already have a roof over your head... So it makes it very difficult for us to really jump up and be able to get something as soon as possible... I've been doing this ever since [my daughter's] been little, and it's very hard."

Balancing Caregiving with Paid Work

Many respondents were employed on a full- or part-time basis in addition to being a caregiver and counted on their income to cover as much of their household's expenses as possible. For most caregivers, employment that was feasible to take on in addition to caregiving had to be local and/or convenient to access en route to institutionalized forms of care like schools. When caregivers had to travel further for work, social support systems for helping with caregiving were crucial.

Many caregivers reported looking for or intending to look for new work that would improve caregiving conditions, such as by providing higher wages, more convenient hours, or less physical toll than their current job. However, for some, it was difficult to find the time to engage in a full job search. Some shared the wish of moving in order to have access to better job opportunities. One respondent shared that she had moved to her current neighborhood in order to be able to support her dependents on a single person's salary. "I had kids and it became harder with jobs and all that," she explained. "So, I ended up moving here, which is a little easier and it goes with my budget." Some respondents faced difficult calculations about whether or not to work a job in order to best provide for their dependents. Since working would mean sacrificing access to certain welfare options and needing to pay for childcare, it was not always the most sensible option.

Some respondents were unable to work, however, whether because of the demands of caregiving or because of other issues that precluded them from doing so. One caregiver explained that she was unable to work because of health issues, but volunteered in order to get a voucher for her grandkids to go to camp during the summer. Predictably, losing a job could be destabilizing, particularly for caregivers with high housing costs. A respondent who worked as a nanny said that she changed jobs often and described a recent time when she was out of work. "I couldn't pay my rent, couldn't pay my bills. And I couldn't afford certain things for my kids…in six months, it will change again because I have to look for a new job because my job is ending. So, I can't survive." Respondents' efforts to balance care with paid work depended on important components of the urban infrastructure of care like accessibility, economic opportunities, childcare and activities, and social support, as well as respondent's ability to coordinate across these different components day-in and day-out.

Summary and Implications for Practice

In order to understand how everyday household and family care work engages and responds to conditions of the urban environment, we analyzed interviews with caregivers across the Boston metropolitan area about their experiences in their neighborhood. Interviews were designed by residents of study neighborhoods as part of a longitudinal PAR project investigating the relationship between urban development and community health. We found that caregivers adopt patterns of strategies that take up or fend off various aspects of the urban environment in order to fulfill their caregiving responsibilities, from the day-to-day routine of putting food on the table to the years-long process of securing suitable and affordable housing. These strategies can be grouped into six domains: managing the dependent's relationship to their environment, securing and maintaining the material conditions for care, creating support systems, preserving health, time management, and balancing care with other forms of work. These strategies comprise major domains of family caregiving and show just how much of everyday household functions actually depend on the broader urban environment past the walls of the home.

Our findings thus show that the urban environment plays a significant and important role in shaping what the work of everyday family caregiving entails for caregivers. If the livability of cities for families, and the potential of cities to promote the health and well-being of families, matter to urban planners and policymakers, then it is imperative that we begin to attend to and account for the ways that cities facilitate or obstruct the strategies caregivers employ to fulfill their caregiving goals. These strategies engage with, and depend to a significant extent on, the affordances of the urban environment. How easy or difficult it is to be a caretaker in a city is very much a reflection of how livable that city actually is, and thus how we measure and assess changes in urban livability should account for how conducive cities are to caregiving. Environmental interventions to enhance family well-being can focus on enabling caregivers to carry out caregiving strategies with greater ease, for example through material improvements in transportation infrastructure, making housing more affordable and making it easier to make affirmative moves as family circumstances change, economic development strategies that prioritize the needs of working caregivers, or community development strategies that seek to increase social cohesion and support. Such interventions can be evaluated according to the degree to which they make the caregiving strategies identified via our analysis easier, less resource intensive, and more fulfilling to carry out (Binet et al., 2022).

Decades of research on urban planning shows that the components of the urban environment that caregivers rely on to carry out caregiving strategies – like housing, transportation, economic opportunities, food access, and public space – are unevenly distributed in cities in ways that both reflect and perpetuate racial and class inequities in society more broadly (e.g., Goetz et al., 2020; Golub et al., 2013; Horst et al., 2017; Manville et al., 2020; Rothstein, 2017; Steil, 2018). This, in turn, suggests that how demanding caregiving strategies that depend on the urban environment are, and how readily caregivers are able to fulfill their caregiving goals by pursuing these strategies, varies along these axes of inequality. Moreover, efforts to address racial and economic inequalities in cities should include efforts to address inequities in the urban environmental affordances, social and material, that enable caregivers to carry out caregiving strategies with ease.

Urban planners and those in allied fields like architecture and design should seek to minimize, as far as possible, the extent to which urban environments make care work burdensome. Doing so promises to alleviate racial, gendered, economic, and health inequities reproduced by the urban environment and enhance family well-being.

References

Arcaya, M. C., Schnake-Mahl, A., Binet, A., Simpson, S., Church, M. S., & Gavin, V. (2018). Community change and resident needs: Designing a participatory action research study in metropolitan Boston. *Health & Place, 52*, 221–230. https://doi.org/10.1016/j.healthplace.2018.05.014

Biglieri, S. (2017). Implementing dementia-friendly land use planning: An evaluation of current literature and financial implications for Greenfield development in suburban Canada. *Planning Practice & Research, 33*(3), 264–290. https://doi.org/10.1080/02697459.2017.1379336

Binet, A. (2021). *Making the city livable: Caregiving and health in gentrifying Boston* [Doctoral dissertation, Massachusetts Institute of Technology]. https://hdl.handle.net/1721.1/140051

Binet, A., Gavin, V., Carroll, L., & Arcaya, M. C. (2019). Designing and facilitating collaborative research design and data analysis workshops: Lessons learned in the healthy neighborhoods study. *International Journal of Environmental Research and Public Health, 16*(3), 324. https://doi.org/10.3390/ijerph16030324

Binet, A., Houston-Read, R., Gavin, V., Baty, C., Abreu, D., Genty, J., Tulloch, A., Reid, A., & Arcaya, M. (2022). The urban infrastructure of care. *Journal of the American Planning Association, 0*(0), 1–13. https://doi.org/10.1080/01944363.2022.2099955

Black, S. (2020). *Social reproduction and the city: Welfare reform, child care, and resistance in neoliberal New York*. University of Georgia Press.

Bondi, L. (2003). A situated practice for (re)situating selves: Trainee counsellors and the promise of counselling. *Environment and Planning A: Economy and Space, 35*(5), 853–870. https://doi.org/10.1068/a35135

Conradson, D. (2003). Spaces of care in the city: The place of a community drop-in centre. *Social & Cultural Geography, 4*(4), 507–525. https://doi.org/10.1080/1464936032000137939

Curran, W. (2017). *Gender and gentrification*. Routledge.

Daepp, M. I. G., Binet, A., Gavin, V., & Arcaya, M. C. (2021). The moving mapper: Participatory action research with big data. *Journal of the American Planning Association, 88*(2), 179–191. https://doi.org/10.1080/01944363.2021.1957704

Daminger, A. (2019). The cognitive dimension of household labor. *American Sociological Review, 84*(4), 609–633. https://doi.org/10.1177/0003122419859007

Duffy, M. (2011). *Making care count: A century of gender, race, and paid care work*. Rutgers University Press. http://muse.jhu.edu/book/2209

Duffy, M., Albelda, R., & Hammonds, C. (2013). Counting care work: The empirical and policy applications of care theory. *Social Problems, 60*(2), 145–167. https://doi.org/10.1525/sp.2013.60.2.145

Folbre, N. (2012). *For love and money: Care provision in the United States*. Russell Sage Foundation. http://muse.jhu.edu/book/16395

Forsyth, A., Molinsky, J., & Kan, H. Y. (2019). Improving housing and neighborhoods for the vulnerable: Older people, small households, urban design, and planning. *Urban Design International, 24*(3), 171–186. https://doi.org/10.1057/s41289-019-00081-x

Fraser, N. (2016). Contradictions of capital and care. *New Left Review, 100*, 99–117. https://newleftreview.org/issues/ii100/articles/nancy-fraser-contradictions-of-capital-and-care

Goetz, E. G., Williams, R. A., & Damiano, A. (2020). Whiteness and urban planning. *Journal of the American Planning Association, 86*(2), 142–156. https://doi.org/10.1080/01944363.2019.1693907

Golub, A., Marcantonio, R. A., & Sanchez, T. W. (2013). Race, space, and struggles for mobility: Transportation impacts on African Americans in Oakland and the east bay. *Urban Geography, 34*(5), 699–728. https://doi.org/10.1080/02723638.2013.778598

Greenwood, D. J., & Levin, M. (2006). *Introduction to action research: Social research for social change* (2nd ed.). Sage.

Hayden, D. (1981). *The grand domestic revolution: A history of feminist designs for American homes, neighborhoods, and cities*. MIT Press.

Horst, M., McClintock, N., & Hoey, L. (2017). The intersection of planning, urban agriculture, and food justice: A review of the literature. *Journal of the American Planning Association, 83*(3), 277–295. https://doi.org/10.1080/01944363.2017.1322914

Kleinman, A. (2015). Care: In search of a health agenda. *The Lancet, 386*(9990), 240–241. https://doi.org/10.1016/S0140-6736(15)61271-5

Lawson, V. (2007). Geographies of care and responsibility. *Annals of the Association of American Geographers, 97*(1), 1–11. https://doi.org/10.1111/j.1467-8306.2007.00520.x

Lefebvre, H. (1996). *Writings on* cities (E. Kofman & E. Lebas, Trans., 1 ed.). Wiley-Blackwell.

Manville, M., Monkkonen, P., & Lens, M. (2020). It's time to end single-family zoning. *Journal of the American Planning Association, 86*(1), 106–112. https://doi.org/10.1080/01944363.2019.1651216

Milligan, C., & Wiles, J. (2010). Landscapes of care. *Progress in Human Geography, 34*(6), 736–754. https://doi.org/10.1177/0309132510364556

Miraftab, F. (2021). Global restructuring of social reproduction and its invisible work in urban revitalization. In L. Peake, E. Koleth, G. Tanyildiz, R. Reddy, & D. Patrick (Eds.), *A feminist urban theory for our time: Rethinking social reproduction and the urban* (pp. 138–161). Wiley.

Mohandesi, S., & Teitelman, E. (2017). Without reserves. In T. Bhattacharya (Ed.), *Social reproduction theory: Remapping class, recentering oppression* (pp. 37–67). Pluto Press.

Parker, B. (2017). *Masculinities and markets: Raced and gendered urban politics in Milwaukee*. University of Georgia Press.

Poo, A-J. (2016). *The age of dignity: Preparing for the elder boom in a changing America* (Reprint ed.). The New Press.

Randles, J. (2020). "Willing to do anything for my kids": Inventive mothering, diapers, and the inequalities of carework. *American Sociological Review, 86*, 35–59. https://doi.org/10.1177/0003122420977480

Roberts, D. (1997). *Killing the black body: Race, reproduction, and the meaning of liberty*. Vintage Books.

Rothstein, R. (2017). *The color of law: A forgotten history of how our government segregated America*. Liveright.

Steil, J. (2018). Antisubordination planning. *Journal of Planning Education and Research, 42*, 9–18. https://doi.org/10.1177/0739456X18815739

Stevano, S., Mezzadri, A., Lombardozzi, L., & Bargawi, H. (2021). Hidden abodes in plain sight: The social reproduction of households and labor in the COVID-19 pandemic. *Feminist Economics, 27*(1–2), 271–287. https://doi.org/10.1080/13545701.2020.1854478

Tarrant, A. (2018). Care in an age of austerity: Men's care responsibilities in low-income families. *Ethics and Social Welfare, 12*(1), 34–48. https://doi.org/10.1080/17496535.2017.1413581

The Care Collective. (2020). *The care manifesto: The politics of interdependence*. Verso.

The Healthy Neighborhoods Study. (2020). *A participatory action research field guide*. https://www.clf.org/publication/field-guide-for-participatory-action-research/

Tronto, J. C. (1993). *Moral boundaries: A political argument for an ethic of care*. Routledge. https://www.routledge.com/Moral-Boundaries-A-Political-Argument-for-an-Ethic-of-Care/Tronto/p/book/9780415906425

Winant, G. (2021). *The next shift: The fall of industry and the rise of health care in Rust Belt America*. Harvard University Press. https://www.hup.harvard.edu/catalog.php?isbn=9780674238091

Part IV
Overview and Next Steps

Chapter 8
Understanding Environmental Impacts on Family Functioning in Service of Resilience and Equity

Devin Malloy McCauley and Jorden E. Jackson

The chapters in this volume provide insights into the many ways in which environmental factors shape the functioning, health, and well-being of families and their members. Our goal in this final chapter is to integrate the diversity of knowledge, perspectives, and research methodologies in this volume to highlight salient themes addressed by the authors – centering the concepts of family resilience and social equity. Toward this goal, we have organized this concluding chapter around three broad questions that emerged from the previous chapters:

(a) How does the built environment shape family relationships and well-being?
(b) How do families respond and adapt to disasters?
(c) How does climate change impact family functioning?

In addressing these questions, we begin from a micro-level perspective, drawing on insights from the family systems framework, and then expand our scope to examine how environmental factors interact with social structures and systems to shape families and children at the population level. We conclude by highlighting key topics to be addressed by future research and prevention efforts related to environmental impacts on families.

D. M. McCauley (✉)
Department of Pediatrics, Stanford University School of Medicine, Stanford, CA, USA
e-mail: dmm16@stanford.edu

J. E. Jackson
Department of Sociology, Penn State, University Park, PA, USA
e-mail: jej5332@psu.edu

S. E. Ortiz et al. (eds.), *Environmental Impacts on Families*, National Symposium on Family Issues 12, https://doi.org/10.1007/978-3-031-22649-6_8

A Family Systems Perspective for Studying Environmental Impacts on Families

The family systems framework offers a number of foundational principles that characterize family functioning and in turn provide insight into the health and well-being of its individual members. Notably, the concept of *wholism* emphasizes that families are systems governed by patterns and rules and that individual behavior is best understood within the broader family context (Minuchin, 1985). Similarly, the principle of *interdependence* highlights the relational and interactive nature of family processes, whereas *circular causality* reveals that family interactions tend to be patterned and reciprocal rather than linear (Minuchin, 1985; Nichols & Everett, 1986).

These principles direct attention to the complexities of family functioning and illuminate the ways in which the development and well-being of individual members are contextualized by broader family patterns and relationships. Therefore, from a family systems perspective, instilling positive change for children and adults requires evaluating and addressing the relational aspects of the family unit as whole, rather than evaluating and treating individuals in isolation (Minuchin, 1974). For this reason, families are salient contexts for both empirical investigation and prevention efforts aimed at supporting healthy human development (e.g., Connell et al. (2015), Kumpfer and Alvarado (2003)).

Though family research and prevention efforts often focus on intrafamily dynamics, the systems framework also emphasizes that families are *open systems* – meaning that family functioning is influenced by interactions with the surrounding environment (Nichols & Everett, 1986). This concept is supported empirically. For example, on days when adolescents experience problems at school, they are more likely to report conflictual interactions with their parents at home (Timmons & Margolin, 2015), and parents' daily stressful experiences in the workplace are linked to later family interactions at home (Repetti & Wang, 2010). Sociocultural factors, such as racial identity, experiences of racial discrimination, and racial socialization, also have implications for couple and parent-child relationships (Jenkins et al., 2020; Lavner et al., 2018).

The concept of families as open systems is particularly valuable when conceptualizing family stress and resilience, as it draws our attention to the risks, resources, and support systems embedded throughout the families' broader environments (Masten, 2021; Witting et al., 2021). As demonstrated throughout this volume, nesting family science within bioecological models of development yields a deeper understanding of the myriad ways in which environments directly and indirectly impact family functioning, bringing to light novel implications for prevention, policy, and future research (Bronfenbrenner & Morris, 2006). Shifting our lens to a macro-level perspective, several chapters in this volume also demonstrate how structural characteristics of environments – including economic, geographic, environmental, and political factors – impact family functioning in ways that account for disparities in health and well-being at the population level. Collectively, the diversity of perspectives and approaches in this volume exemplifies the ways in which strengthening an interdisciplinary approach to family science can enhance understanding of family risk and resilience and also inform and guide efforts aimed at promoting greater equity in our societies. In an ever-evolving world with growing wealth inequality, persistent housing insecurity, changing environmental conditions, and more frequent and intense natural and human-made disasters, the need for interdisciplinary study of environmental impacts on family functioning becomes all the more pressing. With this in mind, we now examine how the chapters in this volume address the role of the built environment in family functioning.

The Built Environment Shapes Family Relationships and Well-Being

Ferguson and Evans (Chap. 5) advance the concept of the family as an open system by drawing upon an extension of Bronfenbrenner's (1977) bioecological model to illuminate the ways in which the physical environment shapes family processes and child's well-being. The bioecological model emphasizes that the home and surrounding areas, such as neighborhoods, parks, and schools, are formative spaces for human development (Bronfenbrenner & Morris, 2006). How, then, might physical surroundings characterized by toxins, pollutants, water insecurity, extreme temperatures, lighting, noise, and crowding impact the family relationships, processes, and routines that guide child development?

The direct impacts of such physical factors on child health are generally well studied, with noted effects on physical, cognitive, and socioemotional well-being (see Evans (2006), Ferguson et al. (2013)). However, the ways in which such factors *indirectly* affect child health through disruptions in family routines and relationships represent a crucial domain for future investigation. In this case, embedding the family systems concept of *interdependence* within a bioecological framework may guide future inquiry. For example, parents working in noisy or chaotic environments are susceptible to fatigue and psychological stress, which may in turn influence parenting practices and family routines in ways that impact child development. Exposure to toxins or lack of access to clean water impacts not only children's physical health (Bartlett, 2003; Ferguson et al., 2013) but may result in increased absences from school – further disrupting opportunities for children's learning, social interactions, and growth (Bartlett, 2003; Eccles & Roeser, 2015).

Notably, Ferguson and Evans (Chap. 5) emphasize that the direct and indirect effects of the physical environment do not act in isolation but rather interact in complex ways to shape family processes and child development. Furthermore, structural forces such as housing segregation and school districting also create vast inequities in the quality of physical environments in and around the home – leaving some populations with greater exposure to multiple physical risk factors (Shonkoff et al., 2021). Therefore, accounting for the cumulative effects of physical environmental factors within and beyond the home is essential for understanding the risks faced by families and children and for designing effective and equitable interventions for improving physical spaces in support of healthy child development.

Binet and Arcaya (Chap. 7) expand upon the bioecological perspective introduced by Ferguson and Evans (Chap. 5) to examine how families navigate the goals and demands of caregiving within modern urban environments. As articulated by the participants in Binet and Arcaya's study, caregiving is not a dyadic act confined to the home but rather relies upon and contends with the networks of private (e.g., home), public (e.g., parks, playgrounds, transportation), and professional (e.g., health clinics) spaces that comprise families' built environments. However, understanding the demands of caregiving in urban settings also requires shifting our focus to the structural forces that guide their development and design. Despite the centrality of caregiving for human relationships and development, many capitalist societies have increasingly relegated the burden of care to families while allocating fewer resources to institutions and social programs that support care. The result is a privatization and commodification of care with ever-widening disparities in who can afford access and who must bear the burdens on their own.

What does this "crisis of care" look like for families? Interviews with participants in the Healthy Neighborhoods Study revealed common ways in which caregiving demands are shaped by constraints of urban environments. Many caregivers discussed the complexity of balancing employment opportunities with accessibility to schools, affordable housing, and other institutional supports. Caregivers also shared their difficulties in managing their dependents' exposure to unsafe spaces in their own neighborhoods while still fostering opportunities for social interactions and growth. Notably, many participants reported having to invest substantial time and effort into creating their own networks of care within their neighborhoods and communities in order to provide for their dependents. In many cases, the constraints of their urban environments made it impossible to meet all caregiving goals at once, leaving families to consistently seek alternative solutions for managing the burdens of care. Unsurprisingly, the unequal distribution of caregiving burdens along racial, gender, and class only perpetuates social inequities along such lines (Daminger, 2019; Duffy, 2011; Folbre, 2012). Therefore, solving this crisis requires investing in structural changes through policy and urban planning efforts that prioritize the demands of caregiving. In other words, building "infrastructures of care" will not only support dyadic caregiving but address broader societal inequities in the burdens of care.

Thus far, these two chapters have enriched our conceptualization of families as open systems. Ferguson and Evans (Chap. 5) illustrate how a bioecological framework can reveal cumulative risk to family functioning within the proximal physical environment, and Binet and Arcaya (Chap. 7) demonstrate that caregiving in urban environments is often constrained by structural forces in ways that perpetuate population-level disparities in family health. In combination, this work illustrates a clear and urgent need for interventions that facilitate change in the built environment in support of family health and well-being. Given this clear need, how can we begin to address the crisis of care and environmental risks to family and child health?

Lachance et al. (Chap. 6) answer this question by providing a framework for supporting community-driven efforts to create lasting, sustainable change in support of child and family health. As they emphasize, the physical health of children and families is dependent on broader social and economic conditions that shape the infrastructure of communities and in turn provide opportunities for physical activity. Efforts to create meaningful change in the built environment are best served by centering equity via community-driven approaches and systems-level thinking. These values were applied by Lachance and colleagues in the Lower Rouge River project, which focused on collaborating with Michigan communities to create safe and equitable opportunities for physical activity along Lower Rouge waterways. For example, assessing the structural determinants of and barriers to physical activity within participating communities ensured that these factors informed the priorities of the project, resulting in *equitable* and safe opportunities for physical health. Adopting a *community-driven* approach meant amplifying the voices of community members and empowering them to author their own vision and goals for the project. *Thinking systemically* required accounting for the interconnected institutions, organizations, and stakeholders within and across these communities and forging coalitions among them in order to build community-wide investment and support for the project. Through adherence to such principles, the Lower Rouge River project was able to empower communities to facilitate structural changes in the built environment that created access to and engagement with the Lower Rouge waterways, in turn promoting equitable and safe opportunities for family physical activity and health.

How Families Respond and Adapt to Disasters

Rapid environmental changes and destruction wrought by disasters carry unique challenges for families and often exacerbate societal and global inequities. Such disasters – natural, technological, human-made, or biological – impact physical and mental health, alter life trajectories, and generally create instability (Abramson, 2021). Though the literature on disasters focuses mainly on individuals and/or communities as the unit of analysis, studying families in the context of disasters provides novel and valuable insights that may facilitate recovery efforts and opportunities for intervention.

During a disaster, families may experience injuries and property destruction, which may in turn precipitate long-term consequences such as housing loss, economic hardships, and separation from community support systems (Abramson, 2021). These acute and chronic stressors collectively increase risk for parents' psychological distress and disrupted parenting – factors that compound risks to children's own recovery and well-being (e.g., Cobham et al. (2016)). Integrating family resilience theories with disaster research provides a framework for examining the interplay of (a) stressors induced by disaster exposure and (b) the family's resources in shaping coping and recovery over time (Figley & Kiser, 2013). Family cohesion, communication, supportiveness, and problem-solving skills during and after a disaster may support individual members' coping and recovery. However, families are embedded within broader social ecologies that carry their own risk and resilience factors, contributing to collective stressors that are not distributed equally among all who are affected by a disaster. Socioeconomic status, access to resources, and community supports shape families' pre-hazard vulnerabilities and play an important role in disaster recovery. Therefore, preparing communities to adapt to environmental disasters may benefit from equity-focused strategies that bolster family and community resilience, while also accounting for immediate disaster-related stressors as well as those that unfold along the road to recovery.

Empirical knowledge of family resilience post-disaster is provided via longitudinal data and natural experiments. Frankenberg et al. (Chap. 1) utilizes one such dataset to predict variation in long-term mental health and mortality outcomes for survivors of the 2004 tsunami in Indonesia. Some communities suffered high death rates, while other communities nearby were relatively unaffected, providing an opportunity to draw comparisons between the long-term outcomes of survivors residing in both types of communities. This study found that communities unaffected by the tsunami had a higher mortality rate at the 15-year follow-up compared to communities hit hard by the tsunami (27% and 23%, respectively). Although perhaps surprising initially, these differences in community-level mortality rate suggest that the tsunami killed mostly frail individuals, leaving the tsunami-affected communities with a more robust population on average compared to the communities unaffected by the tsunami. On an individual level, Frankenberg and colleagues (Chap. 1) show how loss of a spouse and/or close kin influences both mortality and post-traumatic stress many years after the disaster. However, these links vary by gender. Regression results disaggregated by age and gender show that older men who lost a spouse had reduced mortality risk while older women who lost a spouse had increased mortality risk. Further, younger women in communities who experienced the death of spouse or close kin displayed long-lasting psychological difficulties five and ten years after the tsunami. Such gender disparities in mortality and mental health highlight the need for attention to the distinctive experiences of women and men and have important implications for disaster-related intervention and policy. Findings also showed that poor housing conditions predicted both mortality and poorer mental health for women and men of all ages, pointing to a need for housing assistance and higher-quality housing for families in the wake of climate disasters. Together the findings highlight the importance of conceptualizing disaster resilience as a process that unfolds across time, including different outcomes based on gender, age, and family role. By altering the structure of families through mortality, housing insecurity, and displacement, disasters continue to impact the functioning of surviving family members long after the initial event.

While disasters have been shown to cause lasting trauma for survivors, some disaster research has provided a foundation for interventions designed to mitigate risk and alleviate symptoms. Powell et al. (Chap. 2) describe one such intervention, which was aimed at reducing psychopathology for adolescents experiencing anxiety, depression, and post-traumatic stress disorder after Hurricane Katrina. The intervention, Journey of Hope, involves eight group sessions wherein youth learn how to process difficult emotions and engage in positive peer interactions. Additionally, two caregiver sessions help caregivers cope with the effects of the disaster and support their children. Research has shown that Journey of Hope is effective at both reducing stress and increasing coping skills for children and caregivers. Thus, the program has been adapted and implemented in many different countries. For example, after the 2011 earthquake in Christchurch, New Zealand, local stakeholders, including staff from Save the Children, New Zealand, psychologists, and social workers adapted the Journey of Hope program to fit the cultural needs of New Zealand children. In this case, American English was changed to New Zealand English, terminology was changed to reflect the New Zealand education system, and books that were used in the USA were substituted for ones used throughout the New Zealand educational curricula. As Journey of Hope continues to provide services to youth and caregivers post-disaster, collecting data on program implementation, including fidelity and acceptability, as well as program effectiveness, including the long-term impacts of program participation, remains a priority.

How Climate Change Impacts Family Functioning

In addition to the impacts of the built environment and disasters, climate change also has distinctive impacts on families. Such impacts are evident in many facets of family life, including but not limited to family members' physical health (Dorélien & Grace, Chap. 3), mental and behavioral health (Billiot, 2021), family migration (Carrico, Chap. 4), and more broadly community cohesion (Billiot, 2021). As with other environmental impacts, climate change interacts with geographic, economic, and cultural factors to produce unique challenges for families – as well as opportunities for adaptation. Understanding the complex intersection of these dynamic factors is imperative for supporting family adaptation in the context of environmental disruptions due to climate change.

Beyond the connections between the environment and long-term mental health outcomes (Frankenberg et al., Chap. 1), changing climate conditions can interact with a community's cultural history to yield complex challenges (Billiot, 2021). For example, indigenous populations in the USA have experienced unique health challenges resulting from the nation's history of colonial discrimination. Despite well-documented health disparities for tribal populations, the effects of exposure to climate change on indigenous health has not been adequately studied. As described by Billiot in her presentation at the 2021 National Symposium on Family Issues, on which this volume is based, indigenous tribal communities living in coastal regions in Louisiana have suffered substantial land loss, which threatens not only their agricultural livelihood but intergenerational transfer of cultural knowledge grounded in these ancestral lands (Billiot et al., 2019). Billiot (2021) described how indigenous sovereignty, tribal approval, indigenous knowledge, and collective community action must be prioritized in research on climate change adaption within indigenous communities. Decolonial methods include seeking participation *with* rather than imposing interventions *upon* indigenous people, eliminating power imbalances between Western knowledge and indigenous knowledge, and rethinking community engagement research to match indigenous worldviews. This work demonstrates the critical intersectionality of place, history, and culture in family-based climate change research. Applying an intersectional framework better positions research to provide richer and more accurate descriptions of social inequalities toward potential directions for addressing the human and social challenges that arise in the context of environmental change and challenge.

Carrico (Chap. 4) examines a different problem related to climate change – the role of families in climate migration. While migration theories agree that family is at the center of migration, few empirical studies have examined climate migration in a family context. For example, in the new economics of labor migration theory, migration is theorized as a family-negotiated coping technique, with the decision to send a migrant serving to diversify family risk and increase family opportunity (Massey et al., 1999; Stark & Bloom, 1985). Other theories, such as the sustainable livelihoods framework (Bebbington, 1999), consider migration as an adaptation to stressors, with household resources impacting the decision to migrate. Carrico (Chap. 4) draws on both frameworks in study of connections between environmental shocks and characteristics of migrant trips. The health of climate migrants and non-climate migrants was roughly equivalent. However, migrants whose trips occurred during or soon after an environmental shock stay abroad for longer periods of time and are more likely to engage in agriculture at their destination. Carrico concluded that while families and households are central components of migration theory, future migration studies should treat families as the unit of analysis and conduct interdisciplinary research involving both family and migration scholars to address the impacts of climate migration on families.

As climate change escalates, innovative family demography research on climate change may be supported by publicly available data resources that capture spatial and temporal nuances. Dorélien and Grace (Chap. 3), for example, provide guidance for researchers on how to link climate change with demographic patterns and cite a number of publicly available datasets that can be linked to climate data. The Demographic and Health Survey (DHS), the Malaria Indicator Survey (MIS), the Multiple Indicator Cluster Surveys (MICS), the Performance Monitoring for Action (PMA), and the World Bank Living Standards Measurement Study – Integrated Surveys on Agriculture (LSMS-ISA) – all include potentially relevant information and allow for such linking capability. In addition, more recently developed datasets now contain the geocodes needed for merging climate data. While demographic data can be high in resolution, however, climate data are often more spatially coarse, leading to coarser data after merging and reduced specificity (see Fig. 3.1, Chap. 3).

Merging demographic and climate data can provide novel insights about population health, migration, and fertility, as well as issues pertaining to spatial and temporal patterns – all of interest to family demographers. Temporal data, such as date of birth, for example, can reveal how individuals exposed to the same climate at given points in time may nonetheless have different health outcomes depending on the life course stage of their exposure. Further, datasets such as the DHS provide data at the individual, household, and spatial cluster levels, allowing researchers to examine multiple levels of potentially influential factors (see Fig. 3.3, Chap. 3). For example, temperature and rain irregularities can impact families in different ways depending on the household's rural or urban status. Urban places often have hotter temperatures compared to rural places, whereas droughts tend to more severely impact rural households.

These data are not without limitations. Dorélien and Grace (Chap. 3) describe how it is sometimes difficult to accurately identify which individuals were exposed to a specific climate event. In addition, because much of the publicly available data focus on health, it is often difficult to measure migration. For example, survey data post-climate exposure does not include information from those who have already migrated and thus will not capture the range of outcomes associated with climate exposures. Another limitation of these data includes age exclusions. For example, many surveys do not include the elderly in their responses, which leaves researchers with incomplete knowledge about aging, migration, and climate change.

Despite limitations, as Dorélien and Grace (Chap. 3) show, these datasets provide opportunities for rich empirical analysis. Using DHS and climate data, for example, Dorélien and Grace documented how environmental changes can impact maternal and child's health. Specifically, their findings revealed that pregnant women's exposure to more hot days was associated with increased risk of stillbirth or miscarriage as well as lower birth weights. In Mali, improved agriculture seasons may help to mitigate these risks by supplying more food during the hunger season and thus increase child birthweight. Future climate research should focus on elderly populations, as well as on couples and households. In addition to expanding scientific knowledge, such research may have important implications for targeting international aid and policy, more generally.

Lessons Learned and Future Directions

What lessons can we learn through an interdisciplinary approach to investigating environmental impacts on family functioning and well-being? We conclude by highlighting two key lessons evident across the chapters in this volume and consider how they may inform future family scholarship.

Thinking Systemically

Integrating family scholarship with research on climate and other components of the physical environment is enriched by systems thinking. Families themselves are complex systems that are guided by patterns and rules, but they are also deeply embedded within broader social, geographic, and political systems. Interdisciplinary efforts to bridge across more micro and more macro theory, data, and methods offer great promise for enriching family science. For example, Carrico (Chap. 4) emphasizes that although climate migrants are often conceptualized as acting individually, their plans, efforts, and decisions are shaped not only by environmental factors but by their relationships and roles within their families. Binet and Arcaya's (Chap. 7) work illustrates how the dyadic relationship between caregiver and dependent is in fact deeply embedded within structural dimensions of the urban environment and broader social policies.

There are many avenues for progressing understanding of family functioning within its environmental context.

First, family scholars should strive to further integrate a family systems framework with a multicultural perspective in order to better understand racial, ethnic, and cultural diversity in family norms and how such norms are shaped by and respond to broader environmental factors and forces. For example, studying the impact of immigration policies, housing opportunities, or disaster response efforts on family functioning requires careful consideration of the culturally specific rules, patterns, histories, and practices that characterize family functioning for the population in question. As described by Billiot (2021), developing effective interventions to support health among an indigenous tribe in Louisiana required understanding the complex intersection between the tribe's history of colonial institutional discrimination, their cultural values, and the environmental conditions affecting their lands throughout Louisiana. Similarly, future research should strive to better understand heterogeneity in family structures, family structure changes and transitions, and how diverse family structures respond to environmental challenges and adapt to climate disasters.

Advancing methodologies in study of environmental impacts on families also offers great promise for future scholarship. As suggested by Ferguson and Evans (Chap. 5), applying intensive longitudinal methods would provide insight into how short-term fluctuations in environmental stressors (e.g., heat exposure, air quality) correspond with changes in family relationships and family members' psychological and physical health on situational or daily timescales (Bolger & Laurenceau, 2013). Pairing these methods with longer-term data collection (e.g., years, decades), as demonstrated by Frankenberg and colleagues (Chap. 1), would progress understanding of the different timescales along which environmental factors shape family risk and resilience. Applying intensive longitudinal methods and long-term data collected at different points in the lifespan would also provide nuanced insights into how environmental factors differentially shape risk and resilience processes as a function of development. Furthermore, as argued by Dorélien and Grace (Chap. 3), improving disciplinary terminology, data linking methods, and spatial resolution are key to understanding how subtle differences in climate and other characteristics of the environment may shape family risks, adaptive strategies, and outcomes.

Community-Driven Family Research and Equity

Several authors in this volume invested in community-based and participatory research. As demonstrated by Binet and Arcaya (Chap. 7) and LaChance and colleagues (Chap. 6), for example, partnering with families and communities in research not only advances scientific discovery but provides a foundation for facilitating positive, long-lasting structural changes to environments. It remains crucial that such efforts are guided by equity principles to ensure that intervention efforts account for the voices, needs, and decisions of community members who are likely to be most impacted. Otherwise, research and intervention efforts run the risk of perpetuating inequities in access to benefits and resources and their associated outcomes that may be afforded by research-driven policies and programs. Notably, an important debate remains as scholars from across disciplines strive to promote equity and positive change while questioning whether to move quickly versus more slowly. As articulated by LaChance and colleagues (Chap. 6), due diligence and investment in systemic changes may take considerable time and effort on the parts of researchers and community members, but in many cases, such time investments may be more likely to yield lasting structural changes in service of equity. There are cases, however, in which interventions cannot be delayed, for example, in the aftermath of climate-induced disasters. Powell and colleagues' (Chap. 2) descriptions of the Journey of Hope demonstrate how a timely and effective intervention can be delivered in support of families' post-disaster recovery and yet exist within a broader partnership framework that facilitates scale-up and contextual adaptations across the globe. In this case, investing in a prevention infrastructure facilitates rapid and culturally relevant responses to climate disasters. For interventions, both fast and slow, striving to center equity when integrating family and environmental science frameworks promises to reveal novel insights about how to empower families and communities facing ever-increasing environmental risks and challenges.

References

Abramson, D. (2021, October 25). Family resilience and disasters [paper presentation]. In *29th Annual National Symposium on Family Issues*. Penn State.

Bartlett, S. (2003). Water, sanitation and urban children: The need to go beyond "improved" provision. *Environment & Urbanization, 15*(2), 57–70. Water, sanitation and urban children: the need to go beyond "improved" provision (sagepub.com)

Bebbington, A. (1999). Capitals and capabilities: A framework for analyzing peasant viability, rural livelihoods and poverty. *World Development, 27*(12), 2021–2044. https://doi.org/10.1016/S0305-750X(99)00104-7

Billiot, S. (2021, October 25). Building community-based adaptations to address health implications of exposure to environmental changes among indigenous peoples [paper presentation]. In *29th Annual National Symposium on Family Issues*. Penn State.

Billiot, S., Kwon, S., & Burnette, C. E. (2019). Repeated disasters and chronic environmental changes impede generational transmission of indigenous knowledge. *Journal of Family Strengths, 19*(1) https://digitalcommons.library.tmc.edu/jfs/vol19/iss1/11

Bolger, N., & Laurenceau, J. P. (2013). *Intensive longitudinal methods: An introduction to diary and experience sampling research*. Guilford Press.

Bronfenbrenner, U. (1977). Toward an experimental ecology of human development. *American Psychologist, 32*(7), 513–531. https://doi.org/10.1037/0003-066X.32.7.513

Bronfenbrenner, U., & Morris, P. A. (2006). The bioecological model of human development. In R. M. Lerner (Ed.), *Handbook of child development* (Vol. 1. *Theoretical models of human development*, 6th ed., pp. 793–828). Wiley.

Cobham, V. E., McDermott, B., Haslam, D., & Sanders, M. R. (2016). The role of parents, parenting and the family environment in children's post-disaster mental health. *Current Psychiatry Reports, 18*(6), 53. https://doi.org/10.1007/s11920-016-0691-4

Connell, A. M., Stormshak, E., Dishion, T., Fosco, G., & Van Ryzin, M. (2015). The family check up and adolescent depression: An examination of treatment responders and non-responders. *Prevention Science, 19*(1), 16–26. https://doi.org/10.1007/2Fs11121-015-0586-3

Daminger, A. (2019). The cognitive dimension of household labor. *American Sociological Review, 84*(4), 609–633. https://doi.org/10.1177/0003122419859007

Duffy, M. (2011). *Making care count: A century of gender, race, and paid care work*. Rutgers University Press. Making Care Count | Rutgers University Press.

Eccles, J. S., & Roeser, R. W. (2015). School and community influences on human development. In M. H. Bornstein & M. E. Lamb (Eds.), *Developmental science: An advanced textbook* (pp. 513–555). Lawrence Erlbaum.

Evans, G. W. (2006). Child development and the physical environment. *Annual Review of Psychology, 57*, 423–451. https://doi.org/10.1146/annurev.psych.57.102904.190057

Ferguson, K. T., Cassells, R. C., MacAllister, J. W., & Evans, G. W. (2013). The physical environment and child development: An international review. *International Journal of Psychology, 48*(4), 437–468. https://dx.doi.org/10.1080%2F00207594.2013.804190

Figley, C., & Kiser, L. (2013). *Helping traumatized families*. Routledge.

Folbre, N. (2012). *For love and money: Care provision in the United States*. Russell Sage Foundation.

Jenkins, A. I., Fredman, S. J., Le, Y., Sun, X., Brick, T. R., Skinner, O. D., & McHale, S. M. (2020). Prospective associations between depressive symptoms and marital satisfaction in black couples. *Journal of Family Psychology, 34*(1), 12–23. https://doi.org/10.1037/fam0000573

Kumpfer, K. L., & Alvarado, R. (2003). Family-strengthening approaches for the prevention of youth problem behaviors. *American Psychologist, 58*(6–7), 457–465. https://doi.org/10.1037/0003-066x.58.6-7.457

Lavner, J. A., Barton, A. W., Bryant, C. M., & Beach, S. R. H. (2018). Racial discrimination and relationship functioning among African American couples. *Journal of Family Psychology, 32*(5), 686–691. https://doi.org/10.1037/fam0000415.supp

Massey, D. S., Arango, J., Hugo, G., Kouaouci, A., & Pellegrino, A. (1999). *Worlds in motion: Understanding international migration at the end of the millennium*. Clarendon Press.

Masten, A. S. (2021). Multisystem resilience: Pathways to an integrated framework. *Research in Human Development, 18*(3), 153–163. https://doi.org/10.1080/15427609.2021.1958604

Minuchin, S. (1974). *Families and family therapy*. Harvard University Press.

Minuchin, P. (1985). Families and individual development: Provocations from the field of family therapy. *Child Development, 56*(2), 289–302. https://doi.org/10.2307/1129720

Nichols, W. C., & Everett, C. A. (1986). *Systemic family therapy: An integrative approach*. Guilford Press.

Repetti, R., & Wang, S. (2010). Parent employment and chaos in the family. In G. W. Evans & T. D. Wachs (Eds.), *Chaos and its influence on children's development: An ecological perspective* (pp. 191–208). American Psychological Association.

Shonkoff, J. P., Slopen, N., & Williams, D. R. (2021). Early childhood adversity, toxic stress, and the impacts of racism on the foundations of health. *Annual Review of Public Health, 42*, 115–134. https://doi.org/10.1146/annurev-publhealth-090419-101940

Stark, O., & Bloom, D. E. (1985). The new economics of labor migration. *The American Economic Review, 75*(2), 173–178.

Timmons, A. C., & Margolin, G. (2015). Family conflict, mood, and adolescents' daily school problems: Moderating roles of internalizing and externalizing symptoms. *Child Development, 86*(1), 241–258. https://doi.org/10.1111/cdev.12300

Witting, A. B., Bagley, L. A., Nelson, K., & Lindsay, T. (2021). Natural disasters and the relational study of the family: A 2-decade scoping review. *International Journal of Disaster Risk Reduction, 52*. https://doi.org/10.1016/j.ijdrr.2020.101990

Index

S. E. Ortiz et al. (eds.), *Environmental Impacts on Families*, National Symposium on Family Issues 12, https://doi.org/10.1007/978-3-031-22649-6

CPSIA information can be obtained
at www.ICGtesting.com
Printed in the USA
LVHW080831080223
738852LV00023B/39

9 783031 226489